The History of Mathematics

# The History of Mathematics

by

Joseph Ehrenfried Hofmann

*Honorary Professor,*
*University of Tübingen*

THE PHILOSOPHICAL LIBRARY
New York

Translated from the German
*Geschichte der Mathematik*
by
FRANK GAYNOR
and
HENRIETTA O. MIDONICK

ISBN 978-0-8065-2922-6

# Contents

## CHAPTER THREE

## CHAPTER FOUR

## CHAPTER FIVE

# Introduction

THE present treatise is intended to convey a comprehensive view of the history of mathematics, with cursory glimpses of the inroads of mathematics into other allied disciplines. The discussion of the mathematics of remote antiquity, a subject given very thorough treatment in other textbooks, is kept very concise here in order to deal in somewhat more detail with the interesting and yet heretofore all too little studied development of mathematics in the Middle Ages and in the Renaissance period.

My most heartfelt thanks are due to my untiring helpers in checking the proofs, Dr. Eugen Löffler, Ministerial Councillor, of Salzburg, *cand. math.* Helmut Salzamann, of Augsburg, and, above all, to my beloved wife. These modest volumes are dedicated to the memory of my dear friend, Heinrich Wieleitner, to whom I owe a debt of gratitude for a great deal of assistance in the field of the history of mathematics.

*Joseph E. Hofmann*

# A General Survey

IT was not suddenly, that mathematical thought arose in the formalistic generalizations of to-day, and in greatly varied configurations according to content; on the contrary, it unfolded over thousands of years in a long, laborious development. In the beginning, there were the prescription-like instructions for calculation and practical directions for the building crafts, in which curious concepts, partly superstitious, partly mystically religious, were associated with inherent practical usability. Systematic points of view of a more general nature appeared among the Pythagoreans (5th century B.C.); clear insight into the nature, the significance and the scope of the method of proof, together with the discovery of irrationals, were to be seen among the Greeks of Southern Italy (about 400 B.C.). At approximately this time, mathematics freed itself from Greek philosophy, to become an independent branch of learning. Nevertheless, because of the nature of the intellectual attitude toward theorizing, its innermost ties with philosophy endured. The inadequacy of its symbolism and its rejection of direct infinitesimal methods notwithstanding, the mathematics of the golden age of Hellenism (3rd century B.C.) presented a self-contained scientific system of great value. It was an abstract intellectual science and it had but little to do with the contemporary natural science which was striving after qualitative explanations on the basis of the external appearance of phenomena; the possibilities of technical application, too, were barely used.

With the incorporation of the kingdoms of the Diadochi into the Roman Empire (Greece in 197 B.C., Syria in 64 B.C., Egypt in 31 B.C.), the Stoa came to the forefront in the domain of world ideology, as did cosmopolitan eclecticism in the field of philosophy. Interest in theoretical mathematics was on the decline, and even the efforts of the neo-Pythagoreans and the neo-Platonists (3rd century, A.D.), to revitalize it, could not rouse it to its previous height. The young Church at first rejected theoretical mathematics as a discipline because of its link to philosophy, as a typical pagan science, and took the attitude of recognizing the admissibility only of the absolutely indispensible practical methods; the last Neo-Platonists summarized the mathematical knowledge accessible to them in comprehensive anthologies and textbooks. At the same time, the first usable beginnings of a symbolic algebra were developing (4th–6th centuries).

The bipartition of 395 A.D. put in jeopardy the cultural unity of the theretofore completely bilingual Empire; as a result of the collapse of the Western Empire (476 A.D.), the West, overrun by Germanic tribes, lost the knowledge of the Greek language, and since the principal writings of Hellenic philosophy and mathematics had not been translated into Latin, a new start had to be made on primitive foundations. In the Eastern Empire, the scientific tradition survived for a long time; Greek science flowed from here to Mesopotamia, Persia and India, where it was combined with existent native fundamentals and led to a system of computation by figures and to the development of trigonometry (beginning in the 5th century). After the Caliphate had been established (Hegira, 622), and placed on a firm footing, the Moslems recorded the mathematical knowledge of the Greeks and Indians in excellent translations (9th century). They carried this knowledge forward skillfully, and, (beginning with the 12th century) passed it on to the Christian West, where, at the same time, Latin translations from the Greek originals were also being completed. The discussion of the Scholastics about the infinite and the continuum (from the 12th century on) and the symbolical and arithmetic demonstration of properties and their changes (from the 14th century on) prepared the ground for the revival of the mathematical interest, directed at first at the translation (15th century), and then the reproduction (16th century) of the Greek originals reemerging piece by piece, and finally at the independent development of the acquired knowledge. At the same time, with the turn toward quantitative physical research and technical utilization, a world picture was formed, based on mathematical-scientific fundamentals. Definitive features were acquired by this world picture through the subsequent transition to infinitesimal methods of analysis, to the concept of function and to a symbolical ideography (17th century).

The unprecedented results of the new procedure, applied from that time on in all fields of exact natural science, led to an extraordinary expansion of the mathematically comprehensible store of knowledge (18th century), but they did also mislead people to undue negligence concerning the methodologically unobjectionable substantiation of the rules to be applied. Only in the 19th century did the wealth of facts previously discovered become clear, only then was it summarized systematically, unified algorithmically and represented with ever more exacting precision, that led to research into the foundations of modern axiomatic thinking and to the turn toward logistics. Then and only then was there made the first serious attempt at regarding mathematics as a historic entity and treating it not only from the biographical point of view, but also according to its contents of problems and ideas, and at trying to understand it in its reciprocal relationship to the other culture-shaping factors of the individual eras.

*From the Early Beginnings to the Age of Fermat and Descartes*

## CHAPTER ONE

# Pre-Greek Mathematics

## 1. PREHISTORY

DEFINITE signs of mathematical skill are encountered wherever the efforts of the students of prehistory have revealed details of the culture of the Stone Age. Structures and burial sites, decorations on weapons and tools, on utensils and jewelry, and first and above all the pieces of crockery, wattlework and woven articles display a sense of form and a high degree of workmanlike familiarity with the fundamental facts of geometry. The poorly preserved inscriptions are mostly still unsolvable enigmas to us. We know at times that the symbols must represent dates or inventories, and we can interpret certain regularly recurrent symbols as numerals. Our remote ancestors were acquainted with the equilateral and isosceles triangle, the square, the rectangle, the circle and the regular hexagon inscribed in it, the sphere, the cylinder, the parallelopiped, the cube, the regular tetrahedron and octahedron, and other such shapes and forms. A fine sense of symmetry is evident in the decorative filling of surfaces with geometric ornaments, arranged either in endlessly continuable strips or in closed, circular configurations.

The simplest numbers were treated as attributes and were regarded occasionally as directly correlated to the objects counted, so that the same identical quantity would be expressed by different symbols (concrete numbers). In general, the method of writing in rows was preferred; higher units were introduced very early, but the systems of gradation did not range too far and there were frequent gaps in it. A certain partiality for doubling and quadrupling was common to the Indo-Germanic races; among the Celts, multiplying by twenty played an important part. Decimal gradations also made their appearance, as did, most notably, the use of groups of five, and the system of expressing numbers by counting up to a higher placed, larger unit. (Examples for all these methods of expressing numbers and for the continued introduction of new individual number symbols, at times even in considerable deviation from the number word itself, can be found among races on high cultural levels and possessing distinct written traditions.) On the whole, a certain, although not too far-reaching, relationship can be detected between Neolithic man and the traditions and customs of the few primitive races which have managed to survive to our days and which obviously lacked the intellectual molding power to become the progenitors of truly civilized people.

Where decipherable written records are accessible to us, we can observe a prescription-like treatment of practical problems of the kind encountered in the determination of boundaries, the building of storehouses, dams, etc. First to appear were the exact rules for square, triangle and parallelopiped; for the other shapes, at first approximations were used; exact rules were only later developments. The use of measures of mass, coins and weights created a need for conversions between a unit and its subdivisions. This laid the groundwork for the introduction of fractions, the simplest ones of which were treated everywhere individually. Also, similar figures led to the recognition of proportions. The slowly developing social and political organization made the use of larger numbers necessary. The changes of the seasons, recognized to be connected in some hard-to-fathom fashion with the changes in the shape of the moon, led to the difficult calendar problem which was approached by the various tribes and states in completely different ways and was in most cases intimately interwoven with religious ideas.

## 2. THE BABYLONIANS (2000–200 B.C.)

Most of the rather numerous known tablets inscribed with texts of mathematical content in cuneiform script originate from the Tigris and Euphrates basin, over a period extending from the second millennium to the second century B.C. Essentially, they represent records of the knowledge of the Sumerians, a race which about 3500 B.C. created an ideographic writing for recording their agglutinated language, consisting chiefly of monosyllabic root words. Later they converted this writing by stylization into a cuneiform script, the basic symbols of which were pressed into soft clay by a three-edged stylus. In their battles against the steadily advancing Akkadians (from 2500 B.C. on), the Sumerians were forced to fall back more and more before the military might of that Semitic race, and they were almost completely deprived of all power by 2000 B.C.; nevertheless they continued to retain cultural leadership. Their script, partly ideographic (as picture writing), partly as syllable signs, was used to write the inflected Akkadian language. The number system of the texts known to us is based on the successive juxtaposition of wedge-shaped ones (𝍫) and hook-shaped tens (𒌋), which were thus used to write the numbers from 1 to 59. The sign 𝍫 was used again to represent the number 60, and the numbers beyond 60 were expressed purely by position, in a sexagesimal system, although the place value of the individual figures was not definitely established. The symbol for an inside gap or "zero" is not well-warranted before about 600. For the fractions ½, ⅓, ⅔ and ⅚, their own number words and individual symbols were at hand. The system probably had its origin in the combination of the earlier, partly pure decimal (Semitic) denotation of mass, weight and coinage, with the duodecimal system. In it, addition was expressed by juxtaposition, subtraction by its own sign, which was also used to represent a number, less than a given rounded off larger number. Text books in astronomy contained data which can be interpreted as the opposition of positive and negative numbers. An abacus was probably used to hold partial results.

In a table of reciprocals, all the mutually complementary pairs of factors of 60 (or a power of 60) were listed and multiplied by the initial values, ½ and ⅔. Later (3rd century) extended tables appeared which were derived, by systematically continued division by 2, 3 and 5, from the standard table (the list of all integral factors below 60 which can be divisors of $60^n$). There were also multiplication tables in which the one-digit numbers, 10 and the exact multiples of 10 below 60 were multiplied by the first twenty numbers and the subsequent multiples of 10; this also served for the multiplication of unit fractions. Since the unit in the position system was not defined, the (exclusively used) computation with submultiple sexagesimal fractions could be carried out completely in integers. Within this range, the Babylonians had a clear understanding of the nature of the common fraction. They had, furthermore, tables of square numbers, of powers, of submultiple square roots and cube roots, and of integral solutions of the equation $x^3 + x^2 = a$. Square roots which were not submultiples were approximated by the simple and repeated application of the arithmetical-geometrical mean,

$$\sqrt{a^2 + B} \approx a + \frac{B}{2a}$$

which was probably established on the basis of auxiliary geometrical considerations. Moreover, there were tables of Pythagorean right triangles constructed in conformity with the conditions, $x = 2uv$, $y = u^2 - v^2$, $z = u^2 + v^2$.

Because of the peculiar ideographic notations, the directions given for the handling of practical tasks are linguistically almost untranslatable and this use of symbols shows a certain kinship with the formulaic algebraic rendering. We find the determination of the volume of the frustum of the square pyramid from

$$h\left\{\left(\frac{a+b}{2}\right)^2 + \frac{1}{3}\left(\frac{a-b}{2}\right)^2\right\},$$

the calculation of the height of the isosceles triangle (the Pythagorean theorem), the approximate value 3 for $\pi$ and the determination of the versed sine in a circle from the diameter and the subtending chord, also the determination of the areas of configurations constructed of circular arcs, which were easily recognizable at sight. The sum of the square numbers formula,

$$\sum_{k=1}^{n} k^2 = \frac{1}{3}(1+2n)\cdot\sum_{k=1}^{n} k$$

was explained by a geometrical consideration, as the summation of arithmetic series was explained in connection with problems of distribution. On other occasions, the Babylonians skillfully solved linear equations of more than one unknown, as well as equations of the type exemplified by $ax + by = C$, $xy = D$, and other equations reducible to this form, for which the transformation $(x + y)^2 = (x - y)^2 + 4xy$ was constantly drawn into use. Cubic problems were reduced by the table for $x^3 + x^2 = a$; even problems of compound interest computation were handled, involving the use of powers of 2. The abundant material gives an interesting view of the formal progress of Babylonian mathematics which exerted in turn a stimulating influence in its contact with the neighboring peoples (Egyptians, Greeks, Indians).

## 3. The Egyptians (2000–500 b.c.)

We are familiar with a great many Egyptian inscriptions written in the hieroglyphic picture-writing (since the 4th millennium b.c.), the symbols of which originally represented words (mostly triconsonantal), but later became mere letter signs and were used for the phonetic complementation of the polyconsonantal words. The nature of the word represented was often indicated by appending a determinative (a sign indicating the meaning of the word). Only the consonants were written, so that we have no knowledge of the sound values of the language, closely connected to the Semitic. The strictly formal hieroglyphic writing, in which calligraphy was of greater importance than orthography, developed (as early as the third millennium B.C.) into the cursive hieratic script, written on papyrus with a split reed dipped in India ink, and finally, as the result of a process of continually advancing abbreviation and simplification, into the demotic writing. The introduction of a calendar, with the division of the year in twelve thirty-day months and five holidays (4241 B.C.), as well as the erection of perfectly executed gigantic structures, such as the pyramids (since approximately 2900 B.C.), are indications not only of a highly advanced technology, but also of sound mathematical foundations. Unfortunately, our knowledge in this respect is limited to a few collections of examples intended for the use of higher-ranking administrative officials. Especially important among these are the *Moscow Papyrus,* the *Leather Roll,* and above all the *Rhind Papyrus,* written by Ahmes in the 17th century B.C. with references to the 19th.

The number system of the hieroglyphic texts appears to be purely decimal, based on juxtaposition; the texts dating from the epoch of the highest political power contain individual symbols for the gradations up to $10^7$, the highest of which disappeared again with recession to less prominent conditions. When counting, the Egyptians used auxiliary "number gestures" which invaded their terminology also; the mentally obtained results were represented on the counting-board which the Egyptians indubitably knew how to use for addition and subtraction as well. The multiplication was done by continued doubling, making use of decupling, too; division was based on the experimentative reversal of the skillfully executed method of multiplication. Of the individual symbols used in the earlier periods for ½, ⅓, ⅔, ¼ and ¾, only those for ½ and ⅔ survived to the later era; only the unit fractions could be directly represented in writing. The other fractions were "silent"; they were represented as sums of unit fractions. For calculation with fractions, certain auxiliary numerals were introduced which had to do with the adoption of common denominators; in this connection, in keeping with the dyadic character of the Egyptian multiplication, use was made of the systematic representation of $\frac{2}{n}$ as a sum of unit fractions in a completely definite, conventional, tabularized form.

The above mentioned collections of examples allude to economic texts where the calculations were performed according to fixed rules; the statement of the result was followed in most cases by a verifying re-calculation. They involved problems requiring linear equations, arithmetical and perhaps also geometrical progressions. The volume of the frustum of the square pyramid was expressed in the form:

$$\frac{h}{3}(a^2 + ab + b^2).$$

The architects made use of the method of representation by projections in making their plans, and they plotted quadratic nets for easier graduation. For the circle the approximation $\frac{\pi}{4} \approx \left(\frac{8}{9}\right)^2$ was traditional, arrived at perhaps through a transformation of one-eighth of the circle into an equivalent right isosceles triangle; the obtained approximate value was used also for the computation of the volume and surface of the cylinder. The likewise occurring Babylonian approximation, $\pi \approx 3$, must have reached Egypt through Palestine; the knowledge of the relationship $3^2 + 4^2 = 5^2$ in the simplest Pythagorean triangle was in all probability an independent accomplishment.

It is questionable whether any comprehensive picture of Egyptian mathematics can be formed on the ground of the all too scant material available. Also the Egyptians mastered the general concept of fractions. Their formal approaches were not developed as skillfully as those of the Babylonians for whom the introduction of an ideographical representation was a necessity dictated by the linguistic shift; the kind of repeated computation for verification, practiced by the Egyptians, aimed at this early date, in the direction of the method of proof.

### 4. THE INDIANS, CHINESE AND MAYAS (3000–500 B.C.)

Our knowledge of the earliest mathematical accomplishments of the Indians is still very incomplete; the excavations at Mohenjo-Daro (Indus Valley—about 3000 B.C.) might be bringing us new information. The first tangible facts are found in the *Sulba-sûtras* (from the 8th century B.C. on)—strictly practical instructions concerning sacral geometry in the construction of sacrificial altars. We do not know the originals, but only annotated editions of a much later era (about 300 B.C) . In addition to the customary contents, they contain also geometrical area transformations by means of the theorem on the completion of a parallelogram, the Pythagorean theorem, examples of rational right triangles, and the combination of an uneven number of equal squares, under application of

$$n = \left(\frac{n+1}{2}\right)^2 - \left(\frac{n-1}{2}\right)^2.$$

Problems of similarity involve pure quadratic equations which are treated in the irrational case, too. Results of the type of

$$\sqrt{2} \approx 1 + \frac{1}{3} + \frac{1}{3 \cdot 4} - \frac{1}{3 \cdot 4 \cdot 34}$$

indicate the nature of the method—the procedure of geometrical-arithmetical mean formation. The squaring of the circle is effected by

$$\sqrt{\frac{\pi}{4}} \approx \frac{7}{8} + \frac{1}{8 \cdot 29} - \frac{1}{8 \cdot 29}\left(\frac{1}{6} - \frac{1}{6 \cdot 8}\right)$$

and the circling of the square by means of

$$\sqrt{\frac{4}{\pi}} \approx \frac{2 + \sqrt{2}}{3}$$

(both restored). In the formulation of the problem there are several features common with the Babylonian, but not in the solution of it. The existence of links to the people of Asia Minor and the Egyptians is still problematical.

Our knowledge of the earliest mathematical achievements of the Chinese is also uncertain. There are records, dating from as far back as about 1100 B.C., mentioning the approximation $\pi \approx 3$, the Pythagorean triangle, 3, 4, 5, the determination of the height from the length of the shadow (the seedling of trigonometry), the treatment of simple motion problems, the false approach to the solution of linear equations, and the distinguishing of several unknowns by colors. The magic square of $3^2$ cells as well as the designation of the eight points of the compass by combination of the signs — (male) and — — (female) into groups of three are part of the symbolism of numbers which attained such a high degree of development in China. The representation of numbers by knotted cords and by the so-called "bamboo numbers" (1–9: I, II, III, IIII, IIIII, T, TT, TTT, TTTT; tens from 10 to 90: ⚊, ⚌, ☰, ☰, ☰, ⊥, ⊥⊥, ⊥⊥⊥, ⊥⊥⊥⊥; hundreds the same as the numbers from 1 to 9) is considered to date from the 6th century B.C.

We know practically nothing about the highly advanced native civilizations that thrived in the Americas in the ancient days; the Mayan system of writing numbers (third millennium B.C.) alone is known today. The numbers below 20 were written by combinations of dots (ones) and bars (fives); the Maya system was positional, with a symbol for 0, and the gradations were 1, 20, $18 \times 20 = 360$, $360 \times 20 = 7200$, etc. The system was related to the division of the year into eighteen 20-day months and five intercalary days, regarded as days of ill luck. This calendar system was completed by a grouping of the days into 13-day weeks. More exact data about the arithmetic of the Mayas are still missing.

The overall situation outlined on the preceding pages indicates the presence of a wealth of theoretical possibilities behind the purely practical methods observed; the Greeks are the race to which we owe an everlasting debt of gratitude for having recognized these possibilities and thereby having made mathematics a true science.

## CHAPTER TWO

# The Greeks
## (about 800 B.C.–600 A.D.)

### 1. THE BEGINNINGS OF MATHEMATICAL THINKING
### (800–400 B.C.)

THE earliest mathematical accomplishments of the Greeks are almost completely unknown to us; it is an established fact that they did not go anywhere beyond the generally current knowledge of their day. The extent of the influence of Egypt (through the Cretan-Mycenaean civilization) and the Near East (through the settlements around the Aegean) is still problematical.

The Herodianic numbers (the decadic units I, Δ, H, X, M, in combination with the quinary units Γ, P, ΓH, etc.) are the initial letters of the corresponding number words. They appear in Attic inscriptions dating from the 6th to 1st centuries B.C., were used juxtapositionally, and served to indicate the columns on the counting-board. From the 5th century B.C. on, also the use of the Milesian letter-numbers becomes demonstrable; in this system, the numbers from 1 to 9, the tens from 10 to 90, and the hundreds from 100 to 900 were represented by the twenty-four standard letters of the Greek alphabet and three older letters (*stigma, koppa, sampi*). Later the thousands were indicated by a low stroke before the letter-numeral, unit fractions by an accent mark after the denominator; moreover, there were individual symbols for ½ and ⅔. Our knowledge concerning the arithmetic utilizing these numbers is derived from occasional examples in the writings of the post-classical period, from discovered papyri, and from 13th and 14th-century Byzantine treatises.

According to the Stoic and Neo-Pythagorean tradition (unreliable and embellished with legendlike features), already Thales the Milesian (624?–548? B.C.) and Pythagoras of Samos (580?–500? B.C.) were supposed to have had considerable special mathematical knowledge. Within the esoteric community of the earlier Pythagoreans, an aristocratically inclined political-religious society (heyday about 500 B.C.), we find beside fanciful speculations on numbers, the first beginnings of a scientific theory of numbers: integers were classified as odd or even, a distinction was made between prime numbers and composite numbers, square numbers were recognized as the sum of consecutive odd numbers, triangular numbers as the sum of consecutive numbers. The unit was not yet regarded as a number, but rather as the source and origin of all the numbers which were produced by the repeated application of the unit.

Scientific mathematics may be considered to have begun with Anaxagoras of Clazomenae (500?–428? B.C.) who asserted that there is no smallest of small quantities and correspondingly, no largest of large quantities. In the fragment of a treatise by Hippocrates of Chios on the quadrature of the lune (440 B.C.?), the method of reaching conclusions already shows a sweeping systematization. Behind the individual statements there seems to lurk—of course, without being clearly stated anywhere—the principle of mean value, according to which a property (in this particular case the quadrability of the lune) established in some special cases, must be considered generally valid. This seems to be in excellent agreement with the understanding of Democritus of Abdera (460?–370 B.C.) who taught the atomism of matter, who discovered the volume of a pyramid and of a cone (perhaps by divisions into layers), but who was nevertheless unable to prove these facts rigorously. Democritus, in whom the passion for things musical, common to all Greeks, became a stimulus to the study of music theory, may have brought into correlation the harmony of a musical interval with the lengths of the parts of a monochord divided by a bridge, the ratio of the parts being expressed in the simplest integers. This view reflects the strongly rationalistically tinted basic attitude of the natural philosophers of those days, which led ultimately to the conviction that the integer was to be regarded as the measure of all things. It was expressed, among other instances, in the *Canon* of Polycleitus of Sicyon (440 B.C.?), and no less clearly in the somewhat later cosmology of Philolaus of Tarentum (died in 390 B.C.?) and the allied theory of the harmony of the spheres, and finally in the belief in the periodical return of all that is alive (transmigration of souls).

## 2. THE ARITHMETICA UNIVERSALIS (ABOUT 400 B.C.)

Among the Pythagoreans of Southern Italy—of whom Archytas of Tarentum (428–365 B.C.) may be regarded as the most important representative—an *arithmetica universalis,* based on a proportion theory involving rational numbers was constructed now, the principal roles in it being played by the three fundamental mean values (the arithmetic, the geometric and the harmonic mean). A certainly not insignificant accomplishment was the symmetrical solution of linear equations of more than one unknown by Thymaridas of Paros (5th century B.C.); quadratic equations were treated by means of completing the square, a method that is suggestive of a Babylonian influence. In the field of planimetry, the theorems involving parallels and angles, the triangle, equality of areas, transformation of areas, the angles of the circle, and similarity were known and systematized to a great extent in their principal features, although they were represented without uniformity or balance as yet. They were first summarized in a textbook-like form by Hippocrates, who taught mathematics for a fee. In addition to

the constructions executed by compasses and ruler, interpolations by means of mechanical geometrical devices with sliding
parts played a big role. They must have applied, also, to the
original construction of the regular five-pointed star, the *pen-
tacle* that was the emblem of the Pythagoreans. Stereometry
was still in its infancy. It began with the purely practical problems of stage decoration (Agatharchus, about 460 B.C.) and was
advanced toward perspective by the writings of Anaxagoras
and Democritus. The regular solids were known from the remote past as ornaments and religious objects, but had not been
studied theoretically as yet.

Anaxagoras already worked on the squaring of the circle
(about 434 B.C.); Antiphon (about 430 B.C.) approximated the
area of the circle (starting with the square) through the systematic construction of inscribed regular polygons of $4 \cdot 2^n$ sides.
Bryson of Heraclea (about 410 B.C.) likewise made use of the
corresponding circumscribed polygons; he must have used the
principle of mean value in endeavoring to prove the existence
of a square equal in area to a given circle. For the division of
angles in general, Hippias, the sophist of Elis (about 420 B.C.)
introduced the curve $\dfrac{y}{a} = \dfrac{\varphi}{\pi/2}$, produced mechanically as the
intersection of a uniformly displaced line with a uniformly
rotating radius vector, and later (about 350 B.C.) utilized by
Dinostratus for the squaring of the circle (hence the name
*quadratrix*). The interpolation of two geometrical means was
reduced by Hippocrates to $\dfrac{a}{x} = \dfrac{x}{y} = \dfrac{y}{b}$ and solved by Archytas
stereometrically by means of the intersection of the cylinder
$x^2 = \xi^2 + \eta^2 = b\xi$ with the cone $y^2 = \xi^2 + \eta^2 + \zeta^2 = b^2\xi^2/a^2$
and the torus $(\xi^2 + \eta^2 + \zeta^2)^2 = b^2(\xi^2 + \eta^2)$.

## 3. About the Irrational (400–325 B.C.)

In connection with the law of the squares in the right triangle (the so-called *Pythagorean theorem*), whose proof on the

basis of many special cases, and whose generalization to include the oblique triangle, must be regarded as accomplishments of the Italian school, a group of right triangles was constructed, whose sides were represented by integers, namely: $n, \dfrac{n^2 - 1}{2}$, $\dfrac{n^2 + 1}{2}$ ($n$ being an odd number). The expectation of being able to "rationalize" each and every right triangle in this fashion proved to be deceptive. One may well approximate the ratio between the diagonal and the side of the square, i.e., $\sqrt{2}:1$, by the fractions $1/1$, $3/2$, $7/5$, $17/12$, etc., which tend to approach $\sqrt{2}$ ever more and more exactly from both sides (the generating numbers can be obtained from the geometrically easily proven identity

$$2x^2 - y^2 = (2x + y)^2 - 2(x + y)^2 = \ldots)$$

but ever greater numbers will be reached without arriving at the exact value. At last (about 400 B.C.), the irrationality of the ratio was recognized, possibly through a reversal of the procedure followed. This discovery is ascribed to Hippasus of Metapontium who is reputed to have been also the discoverer of the construction of the regular dodecahedron. Plato records that Theodorus of Cyrene demonstrated to him, the incommensurability of the irrationals from $\sqrt{3}$ to $\sqrt{17}$, by geometrical reasoning, wherein, perhaps, the impossibility of expressing an irreducible ratio in terms of numbers smaller than the original ones, came into play. The discovery of the process of alternate elimination (chain division) must have been decisive in determining the factors common to given numbers. Mathematicians now made themselves masters of the indirect method of reasoning which Zeno of Elea (490?–430? B.C.) had already used, through his ingenious fallacies to challenge the current views of his day on the nature of space, of time and, chiefly, of motion.

With the demonstration of the existence of irrational linear ratios, the *arithmetica universalis* was shattered. The brilliant

Eudoxus of Cnidus (408?–355? B.C.) nevertheless succeeded (about 370), while giving added depth to the ideas of Anaxagoras, in constructing an ingenious geometrical theory of proportion in which the old one was retained as a special case. He started out from the axiom of mensuration (two magnitudes can enter into an interrelationship solely if by multiplication the smaller of the two can be made bigger than the other one) and determined the sameness of two ratios, $a:b$ and $c:d$, indirectly, in that he required that if $m$ and $n$ are any two numbers prime to each other, the relationship $ma \gtreqless nb$ will always be an indication that $mc \gtreqless nd$ as well. A similar process of reaching a conclusion was applied by Eudoxus also where inadequate infinitesimal considerations had been the only method that enabled the earlier mathematicians to accomplish their purpose. Let the proportionality of the area of the circle, $f$, to the square of the diameter, $d^2$ (recorded in Euclid's *Elements*, XII, 2), be cited as a typical example. Let two homologous circles be designated by $f$ and $f'$; if we inscribe into them the mutually similar regular polygons $f_n < f$, $f'_n < f'$, respectively, we find that $f_n : f'_n = d^2 : d'^2$. If $f$ were not proportional to $d^2$, but, e.g., $f_n : f'_n = (f - \varepsilon) : f'$, ($\varepsilon$ designating an area not in excess of $f_4$), it would be possible to arrive by sufficient doubling of the number of the angles, at $f_n > f - \varepsilon$, and thus $f'_n > f'$, contrary to the original construction. Similarly one can also prove the assumption $f_n : f'_n = f : (f' - \varepsilon')$ false; thus there remains only the possibility $f : f' = f_n : f'_n = d^2 : d'^2$. Eudoxus applied the identical procedure to demonstrate the proportionality of the volume of the sphere to the cube of the diameter, and the correctness of Democritus' determination of the volumes of the pyramid and the cone. For the interpolation of two geometric means he makes use of two higher curves (not definable more closely) and perhaps of the mechanical-geometrical construction using two right angles which was handed down to posterity as "Platonic." The problem of apparent planetary motion led him to the intersecting curve of a cylinder, $x^2 + y^2 = ax$, with a tangent sphere, $x^2 + y^2 + z^2 = bx$. He is said to have been the author of the first textbook on stereometry.

Plato (427–347? B.C.) took an active part in this entire development, at least after his first journey to Italy and Sicily (388–87 B.C.) which brought him in personal contact with Archytas. He handed the theory of the irrational over to Theaetetus (410?–368 B.C.) who carefully studied in minute detail the composite quadratic irrationals and applied them to the relations among the five regular solids and the radii of their respective circumscribed spheres (about 380 B.C.). Plato was by no means a professional mathematician, and the individual achievements accredited to him—as the construction of integral right triangles using $2n$, $n^2 - 1$, $n^2 + 1$, (where $n$ is even) —are unimportant for the overall development. Of decisive importance was, however, his profound fondness for mathematical methods of reasoning; this fondness dictated his requirement that an adequate general mathematical preparatory training precede the study of philosophy proper. Plato was aware of the inadequacies of the previous structure of mathematics and strove for a rigorous, logical system which led him ultimately to the formulation of definitions and axioms, to the elaboration of the direct and indirect methods of reasoning, and to insisting on careful distinction of cases in construction problems. He is said also to have suggested that the construction tools be restricted to the compasses and the ruler. The (lost) elements compiled by Leo (about 370 B.C.) came into existence under his influence; the (likewise lost) revised version of Theudius of Magnesia (about 340 B.C.) extolled the skillful compilation of numerous individual questions.

The ranks of the Platonists included also the brothers Menaechmus and Dinostratus (about 350 B.C.), chiefly disciples of Eudoxus in matters mathematical. Menaechmus was the discoverer of the equilateral hyperbola and of the parabola, which he introduced as "solid" geometrical loci (in contra-distinction to the "plane" loci—the straight line and the circle), on the basis of the equations $xy = ab$ and $y^2 = bx$, and which he used for the interpolation of two geometrical means. Conic sections were first introduced as plane sections of a circular cone (perpendicular to a generating line, so that one might speak of the sections of an acute, of a right, and of an obtuse cone) by the somewhat younger Aristaeus (about 330 B.C.), author of a voluminous treatise on solid loci and of another one on the regular solids (both lost).

Like Plato, also Aristotle of Stagira (384–322 B.C.) was interested in the basic problems of mathematics, but he had less in common with the current research of his time. The proof of the irrationality of $\sqrt{2}$, so frequently mentioned by him, achieved by the contradiction of odd and even numbers, may have been his own accomplishment. His logic was based entirely on mathematical reasoning. A very large part of our present-day knowledge of pre-Euclidian mathematics is based on the numerous data recorded by Aristotle concerning the elementary mathematics of his age, as well as on the more detailed specific comments attached to them by the later commentators, of whom special mention is due in particular to Alexander of Aphrodisias (about 200 A.D.), Porphyry of Tyrus (232?–300? A.D.), Themistius (317?–387 A.D.), Joh. Philoponus (6th century A.D.), and Simplicius (about 530 A.D.). The comment of Aristotle that a straight line can not, on the basis of fundamental principles, be made equal to a curved line, was developed by his commentators, of whom mention is due above all to Averroës (1126–1198 A.D.), into a regular dogma that was still predominant as late as the 17th century and made curve rectification in general appear impossible.

To the disciples of Aristotle the interest in mathematics was subordinate to the encyclopedic interest, although Eudemus of Rhodes did produce a history of mathematics (about 335 B.C.), of which unfortunately only fragments have survived.

### 4. EUCLID OF ALEXANDRIA (365?–300? B.C.)

Euclid was the brilliant crowning glory of the entire process of development before him; moreover, the great Alexandrian introduced a new period in the domain of mathematics. The *Elements* (325 B.C.?)—to be regarded not as some elementary introductory text, but as a curriculum for mature students— reveal the masterly skill of the systematist trained on Aristotelian logic and yet standing entirely on the soil of Platonic ideology. The material is presented in thirteen books, of which the first six deal with planimetry, the next three with elementary number theory, the 10th with the theory of irrationals, and the other three with stereometry. The subject is discussed purely theoretically and without reference to the practical possibilities of computation or application. The presentation begins with definitions, axioms and postulates, and then follow the propositions and constructions, without a single word about the intrinsic coherence of the individual parts or the basic intent behind it all. The text available to us today is based on manuscripts which were produced by no means any earlier than the 7th century A.D. and are certain to have undergone numerous alterations in the text as against the original. Theon of Alexandria (about 370 A.D.) already found it necessary thoroughly to re-edit his greatly impaired copies. That version was followed by all the known Greek copies, up to one rediscovered by Fr. Peyrard (1760–1822) in 1808; the pre-Theonic edition was, however, known to the Arabs, too.

The introductory definitions of Book I (as, e.g., a point is that which cannot be divided) are descriptive and comprehensible solely in connection with the contemporary philosophical discussions. Most of the others are conditions for existence, given as verbal definitions. The postulates were probably an important methodical supplement by Euclid himself. The first three limit the field of construction to the use of compasses and ruler; the fourth one, on the equality of all right angles, and the fifth one, the famous "parallel postulate," were subjects of dispute even in classical antiquity (since Geminus of Rhodes, about 75 B.C., Ptolemy, about 150 A.D., and Proclus, about 450 A.D.). Among the axioms, the one concerning the whole and the part was destined to acquire significance later, in connection with infinitesimal observations (the question of the angle of contingence). In the theorems which contained various implied intuitive elements, all reference to motion was avoided as much as possible (under the influence of Zeno's criticism).

Euclid began with the theory of the triangle, passed on to the theory of parallels (making use here for the first time of the parallel postulate); he proceeded to the comparison of areas and the completion of a parallelogram theorem, and closed with the Pythagorean theorem and its converse. The factual content of these theorems had already been contributed by the Pythagoreans; in Euclid we have the origin of sequential arrangement, systematization, and general procedure for proof of the law of the right triangle, based on the squares upon its sides. The best of what classical antiquity had to say on this subject is in the thorough commentary by Proclus.

Book II contains algebraic transformations, like the calculation of $a(b + c)$ or $(a + b)^2$ in geometrical garb. This serves to bring out the solvability of quadratic equations in general, demonstrated by the example, $x^2 = a(a - x)$. The extended theorem of Pythagoras and the altitude theorem follow. In Book III, the theory of the circle is discussed in a very adroit fashion; deserving of special mention here are the inscribed angle theorem which is formulated and proved as an area theorem, chord, secant and tangent theorems, the assertion later heatedly disputed, that the angle of contingence (between an arc and a tangent) is smaller than any possible acute angle formed by straight lines. In Book IV, there are constructions for the simplest regular polygons in connection with the circle, then the regular pentagon on the basis of the constructions in the second book, and lastly, the regular polygon of 15 sides. This was all Pythagorean, and partly pre-Greek common knowledge. It is treated purely geometrically and without the application of proportion, for which the important Book V is reserved. That book presents the Eudoxus theory of numbers, in all likelihood closely following the original (cumbersome phraseology); the final part is devoted to the treatment of compound ratios (multiplication of fractions). Book VI contains the application to the theory of similarity of the older school which could now be presented perfectly and homogeneously. Of particular importance is the treatment of quadratic equations by the application of areas (in continuation of II, 4/6) which was approached and newly interpreted by Apollonius (about 200 B.C.). This method led to the determination of the largest parallelogram among all those of equal perimeter, having the same identical angles (VI, 27).

Books VII–IX contain the simplest fundamental facts of the Pythagorean number theory, in particular the so-called "Euclid's algorithm" of alternating division for ascertaining the greatest common factor, and then, the least common multiple. To this was added the proof of the uniqueness of factorization into prime factors (IX, 14), and methods of calculation, to the second step, of powers and roots (VIII), the summation of finite geometrical progressions (IX, 35), and the proof of the existence (already known to Plato) of infinitely many prime numbers (IX, 20). Next, the early Pythagorean theory of odd and even numbers was presented. In IX, 36 we find the law of the formation of perfect even numbers (as $6 = 1 + 2 + 3 =$ sum of the factors): If $s_n = 2^n - 1$ is a prime number, $2^{n-1}s_n$ is perfect. The voluminous Book X is based on the previous studies of Theaetetus; it embarks on a broad investigation, difficult to view as a comprehensive whole because of the impractical terminology, with the ultimate purpose of denoting the type of irrational number occurring in the regular solids and ascertaining under what circumstances the symbols for square root in $\sqrt{a + \sqrt{b}}$ may be dispensed with. As we gather from the commentary of Pappus (about 320 A.D.), Apollonius seems to have made no substantial progress in the study of the extended expressions $\sqrt{a + \sqrt{b} + \sqrt{c}}$. Among the introductory general propositions of Euclid, X, 2 is important, viz.: Magnitudes are incommensurable if the Euclidean algorithm applied to them never ends. A lemma to X, 29 contains the general representation of rational right triangles by means of $2mn : (m^2 - n^2) : (m^2 + n^2)$.

Book XI presents some data involving the principle of duality with regard to straight lines and planes, next, the most important theorems on solid angles, and finally, the theory of parallel surfaces. Book XII begins with the determination of volumes, continues with the proportionality of the circle to the square of the diameter (XII, 2) and the proportionality of the sphere to the cube of the diameter (XII, 18), then the relation between the cylinder and the inscribed right cone (XII, 10), all of which are proven in the Eudoxian manner, but established anew in each individual case. An independent achievement of Euclid's is the determination of the volume of the pyramid (XII, 3–5) through a continued subdivision into pairs of prisms which—properly arranged—form a geometrical progression. Book XIII is based on the studies of Eudoxus on continuous division, and on the investigations of Theaetetus concerning regular solids and the irrationalities determined by them; this book concludes with the corollary to these theorems: that there are exactly five regular solids.

Many a rough spot and imperfection notwithstanding, Euclid's *Elements* are a first-rate masterpiece which completely supplanted the former elementary compendia, and which in its basic ideas remained substantially unaltered until modern inquiry into fundamental principles decidedly complemented it. The *Elements* are supplemented by the *Data* (complementary planimetrical theorems as an introduction to the elaboration of an expedient analysis), the *Porisms* (lost theorems, possibly of metric-projective contents), *De Divisionibus* which was a treatise on the division of plane figures (partly preserved in Arabic), the *Fallacies* (lost), and the *Phaenomena*, a *Sphaerica* (geometry of the sphere) which made use also of the kindred researches of Eudoxus (lost) and Autolycus (about 330 B.C.); a widely used revised version was prepared by Theodosius (about 100 B.C.). The *Surface Loci* (curves on the cylinder and on the cone?) has been lost, and so have also the four volumes of *Conic Sections* in which presumably the studies of Aristaeus were continued. In *Optics* we find several facts concerning geodesy and perspective.

### 5. ARCHIMEDES OF SYRACUSE (287?–212 B.C.)

Archimedes, who at a more mature age was a student of the followers of Euclid, left no comprehensive book behind, merely individual treatises on mathematical and mechanical subjects, known to us in the edition, partly also with the commentaries, of Eutocius (about 520 A.D.). The earliest ones were dedicated to the Alexandrian astronomer Conon of Samos (died about 240 B.C.), who had written about conic sections, and others were addressed to Conon's disciple Dositheus (about 240 B.C.), the latest ones to Eratosthenes of Cyrene (276?–194? B.C.) who in 235 B.C. was appointed chief of the Alexandrian library and was a very versatile scholar; in the domain of mathematics, he is credited, in addition to the screening ("sieve") method of the determination of prime numbers, also with a mechanical construction for the interpolation of two geometrical means.

Starting out from studies (now lost) concerning the lever, Archimedes went on to a mechanical and geometric *quadrature of the parabola* (the latter with the aid of the infinite geometrical progression $\frac{1}{4} + \left(\frac{1}{4}\right)^2 + \cdots$), determined the center of gravity of the triangle, pyramid, parabola and paraboloid of revolution, he correlated the volume and surface of the sphere and its parts with the corresponding magnitudes on the circumscribed cylinder, defined the circle approximation $3\frac{10}{71} < \pi < 3\frac{10}{70}$ from the inscribed and circumscribed 96-sided polygon, and he studied the *spiral* $\frac{r}{a} = \frac{\varphi}{a} = \frac{t}{T}$ (which bears his name) from the aspect of its mechanical generation. He proved the area propositions, and in connection with this he summed the series of consecutive square numbers. He constructed the tangent to the spiral, correlating the subtangent with the extension of the arc of the circle. The book on the quadric surfaces of revolution contains many interesting theorems on conic sections, including the quadrature of the ellipse; the actual goal was the cubature of quadric surfaces and their parts. This aim was accomplished by the always newly applied and refined Eudoxian method, which was adequate for proving results already known, but not for finding them expediently.

Just how Archimedes arrived at his new laws and theorems is revealed by him in his *Methodology*, rediscovered by J. L. Heiberg in 1906; it leans on auxiliary considerations of mechanical character and differs from modern integral calculus in matters of form only. It presents the most important ones of the previously achieved individual results, also the determination of the volume of the ungula (cylindrical "hoof") and of the solid core between two cylinders inscribed in a cube. Arabic versions tell us of the assumed theorems which contain the quadrature of figures bounded by circular arcs and the paper strip construction for the trisection of an angle, in addition, of the construction of the *regular heptagon* which is reduced to a surprising case of equal areas. On this occasion, Archimedes presents what is customarily referred to as Heron's formula, and the axiom named after him—a relationship in the circle which, as to contents, corresponds to the Ptolemaic theorem and permits the construction of chord trigonometry, although a little less detailed than through Ptolemy's theorem.

The treatise on the *semi-regular solids* is unknown to us except for title; on the other hand, a monograph on the endlessness of the number system (referred to as the "Sand-Reckoner") has been preserved. The starting point was the heliocentric hypothesis of Aristarchus (310?–230 B.C.), which, however, was then rejected for physical reasons. Let us mention just quite cursorily the treatise (lost) of Archimedes on the burning mirror and his (preserved) monograph *On Floating Bodies* in which the principle named after him is utilized, and furthermore a geometrical game bearing his name and the "cattle problem" which calls for the determination of eight unknowns out of seven equations with additional complicating conditions.

In *Measurement of the Circle,* Archimedes at first used very exact approximations for $\sqrt{3}$, all of which can be explained on the basis of the inequality

$$\sqrt{\frac{a-x}{a+x}} < \frac{2a-x}{2a+x}$$

which is easily represented geometrically; for further calculation he used the Babylonian approximation, $\sqrt{a^2+B} < a + B/2a$. The situation concerning the refinement of the Archimedean circle measurement referred to by Heron (about 100 A.D.), which is supposed to have yielded $\frac{35.312}{67.441} < \frac{\pi}{6} < \frac{32.647}{62.351}$, has not been clarified. The first book *On the Sphere and Cylinder* contains the postulate that the straight line is the shortest distance between two points, and that in convex configurations, the enveloping figure is always larger than the enveloped figure—the condition that until the late 19th century was regarded as the fundamental premise for all rectifications.

Although Archimedes was universally admired and was regarded even in his own time as unquestionably the foremost mathematician of classical antiquity, he had no immediate suc-

cessors. His methods were adopted by his zealous Arabic translators, and were developed further and elaborated into modern
analysis by the great systematizers of the Baroque.

## 6. APOLLONIUS OF PERGA (262?–190? B.C.)

Apollonius was almost a contemporary of Archimedes, and
like the latter, came from the Alexandrian school, where later
he taught. His chef d'oeuvre, the *Conica,* originated in Alexandria, but was finished only in Pergamon where he spent his
manhood. The first four books of *Conica* have been preserved
in Greek, three more in Arabic, and the contents of the eighth
one are known from extensive information recorded by Pappus. The results recorded in the first four books, to which
Eutocius added a commentary, are mostly those of the predecessors of Apollonius, namely Menaechmus (about 350 B.C.),
Aristaeus (about 330 B.C.), Euclid (about 320 B.C.), Conon (died
about 240 B.C.), Archimedes (about 240 B.C.), and others; what
is new, however, is the unifying method. This was the first time
that the conic sections were produced as sections on the same
identical (right or oblique) circular cone and discussed on the
basis of the apparent geometrical relationships which, to-day,
we would infer from the equation in rectangular co-ordinates:

$y^2 = rx \left( 1 \pm \dfrac{x}{t} \right)$, referred to the vertex as origin (where $r =$
*latus rectum,* and $t =$ *latus transversum*). The technical terms
*ellipse, parabola* and *hyperbola* (consisting of two mutually
complementary "counter-sections"), linked to the equivalent
Euclidean problem in the application of areas (*Elements* VI),
make their appearance then. Thereupon follow the theorems
on conjugate diameters, tangents, hyperbolic asymptotes, the
relations between pole and polar, and (in extension of the
theorems on chords and secants of a circle) a great many theorems on area. Then there is the generation of projective relations out of two systems of rays, the study of the properties of
the focal point in central conics, and of the points of intersec-

tion of two conics with concomitant consideration of common points of contact (numerous case distinctions).

The interesting Book V is devoted to the theory of the normals, which are, however, introduced not as lines perpendicular to tangents ("tangent normals"), but as the shortest distances of a moving point of a curve, from a fixed point. There appears in this connection the constancy of the parabolic subnormals (already used by Archimedes). Apollonius formulates propositions which lead him close to the theory of evolutes of the conic sections. In Book VI, congruent and similar conic sections are discussed and fitted as sections into a given cone; in Book VII, numerous special properties of the conjugate diameters are deduced. Book VIII contains problems in construction from given parts (without assumption of the drawn conic section) with careful determination.

Of the further writings of Apollonius, only the one on *Proportional Sections* has been preserved (in Arabic translation); the others are mostly known to us only from the allusions by Pappus (about 320 A.D.). The *Proportional Section,* the *Solid Section* and the *Determinate Section* deal, essentially, with the properties of projective series of points, the *Interpolations* with the part problems of this type, constructible with compasses and ruler, in cases of rectilinear guide curves. In *Plane Loci,* rectilinear and circular loci are discussed, with propositions fundamentally out of the inversion theory already making their appearance. Here, the circle which was named after Apollonius, although it had already appeared in Aristotle's work, was mentioned. In "Contacts," circles that were tangent to three given circles, or their degenerates in the form of points or straight lines, were constructed by means of compasses and ruler. These treatises later engendered the greatest interest in mathematicians of the early Baroque.

We know from further references that Apollonius treated the interpolation of two geometric means by a mechanical-geometrical apparatus and by the application of conic sections.

In the *"Rapid Calculator,"* he strengthened and surpassed
sand reckoning and circle measurement; in the investigation
of the helix, he surpassed the treatment of spirals, in his *Burning Mirror,* Archimedes' study of the same title. Furthermore,
he is to be credited with numerous contributions, elementary
as to content. For example, in the work on regular solids, he
demonstrated that the surfaces of the regular dodecahdron and
icosahedron inscribed in the same sphere are to each other as
their respective volumes. This treatise, in a version edited by
Hypsicles of Alexandria (about 180 B.C.), was appended to
Euclid's *Elements* as Book XIV.

Apollonius was responsible for important accomplishments
in astronomy, too. Like Archimedes, he rejected the heliocentric hypothesis of Aristarchus of Samos (310?–230? B.C.)
which was adopted only by Seleucus, the Chaldean of Seleucia
(about 150 B.C.). He was the creator of the epicycle theory
which was elaborated further by Ptolemy (about 150 A.D.).

All the things that we read about the accomplishments of his
contemporaries and of the next generation (about 180 B.C.)—
the cissoid of Diocles, the conchoid of Nicomedes (connected
with the interpolation problems), the cubature of the torus by
Dionysodorus and the study by Perseus of its plane sections,
the treatise by an unknown author on spherical helices, and
the treatise by Zenodorus on isoperimetric figures—display the
effects of the influence of Archimedes and Apollonius as to
general objectives, but are no longer on the same level as to
details. On the whole, the scholar of Perga became appreciated
again by the Moslems only; in the course of the wrangle about
the methods of Apollonius, the early 17th century was led to
coordinate geometry.

### 7. The Decline of Hellenism; the Romans (150 B.C.–150 A.D.)

We know but very few positive facts concerning the subsequent centuries with their continual political troubles and

squabbles. During that period, in connection with astronom-ical studies, a chord trigonometry came into existence. Aristar-chus (about 270 B.C.) already had known that the ratio between chord and arc in the same circle increased with the decrease in the size of the arc (smaller than a semicircle); the addition theorems for the angle functions can be obtained by means of the axiom of Archimedes almost as simply as with Ptolemy's theorem. The latter had probably been expounded already by Hipparchus of Nicaea (180?–125? B.C.) who conducted astro-nomical observations in Rhodes and Alexandria between 161 and 124 B.C. In elaboration of fundamental data established by the Babylonians, he computed a table of chords (lost to us), utilizing the sexagesimal system of notation; he was in all like-lihood already familiar with the simplest problems of practical trigonometry as well. Moreover, the Moslems ascribe to him an algebraic script, applied perhaps to the treatment of quadratic equations.

Heron of Alexandria is to be assumed to have lived and worked later than Hipparchus (about 100 A.D.); he was an out-standing mechanic and technician whose writings and lectures circulated in numerous editions and versions. Heron com-bined the teachings of the old Egyptian tradition adroitly with the achievements of Archimedes, whom he cited frequently. His theoretical writings seem to have included a Euclid com-mentary; among his contributions intended for practical workers there were calculations of areas and solids—some ac-cording to exact formulas, others by good approximations—including what is known as Heron's formula for the triangle, with proof, a brief outline of the Archimedean methodology, anticipating the so-called Cavalieri's Principle (1635), the method of the arithmetical-geometrical means for the approxi-mation of square roots, and a corresponding method for cube roots. In the treatment of quadratic equations, there was for the first time—perhaps under the influence of Hipparchus—reference to the twofold character of the solutions (only the positive value was allowed). Heron was well taught in the domain of land surveying. The instrument—the *dioptra*—which he used was an excellent implement. The (very poor) mathematical writings of the Roman land surveyors, the *agrimensors* (about 100 A.D.), were based almost exclusively on Heron.

Trigonometry was developed further by Menelaus of Alexandria (about 100 A.D.), who treated primarily of spherical triangles, adding, by means of an intersecting great circle, the "rule of six quantities" (*regula sex quantitatum*) bearing his name. Ptolemy (Claudius Ptolemaeus) of Alexandria (85?–165? A.D.) presented the epicycle theory of Apollonius in an expanded and improved form in his book which the Moslems have called *Almagest** and which for almost fifteen centuries remained an exemplary presentation of the geocentric system. The trigonometric details set forth by him were by no means new as to meaning and content, but were presented with great skill in a methodically precise form. The table of chords computed by him (at intervals of half-degrees) was based on the sexagesimal subdivision of the radius and organized for the interpolation of the minute values. The *Analemma* contains the orthographic projection, previously utilized even by the ancient Egyptians, in the treatment of astronomical problems, and the *Planisphaerium* contains the stereographic projection which may have been originated by Hipparchus; the astrological *Tetrabiblos* can be traced contentwise back to Poseidonius of Apamea (135?–44? B.C.). This was the subject that, more than any other one, attracted the ardent interest of the Indian, Arabic and medieval translators.

The Romans produced no original mathematical accomplishment. Their numerical notation system, with the decimal units I, X, C, M, and the symbols, V, L, D, representing

---

* The Greek title of the book, ἡ μεγάλη σύνταξις: (The Great Collection), became ὁ μέγιστός (complete: βίβλος = book), and, by the prefix of the Arabic article *al*, the word *Almagest.*

multiples of five, obtained by halving the decimal units, was early Italian. Peculiar to it is the method of counting backward from a round figure, applied both in the nomenclature of numbers and in calendrical terminology, even though it was not congruent with the written numerals. The fraction system, rising out of the division of the *as* (a coin) into 12 *unciae,* was limited to duodecimally determinable subdivisions, for which number words displaying perceptible marks of Greco-Southern Italian influence and forms, presumably the products of linguistic erosion, were incorporated into the language. All the rest was just roughly approximated. Even every-day counting necessitated the use of the abacus and reference tables, and was looked upon as very intricate. This primitive system was very tenacious and survived even the era of the great migration of peoples. The purely practical-minded Romans handled only problems relating to the division of legacies, computation of interest and business accounts. They were ignorant of the very rudiments of geometry, so that for the great survey of the territory of the Empire under Augustus (about 30 B.C.) they had to resort to the services of Alexandrian workers.

## 8. NEO-PYTHAGOREANS AND NEO-PLATONISTS (150–600 A.D.)

It was only in the later period of the Empire—when the leadership slipped out of the hands of the true Italian stock, and the Empire under capable rulers, fused into a bilingual cultural unit—that Greek learning gradually experienced a revival, first of all among the Neo-Pythagoreans (100 B.C.–200 A.D.) who, to be sure, were entangled to a great degree in unfounded superstitions and fell victims to alleged miracle-workers like Apollonius of Tyana (1st century A.D.). Only in this period did that Pythagorean legend develop which makes it so difficult for us to form a clear picture of the beginnings of Greek science. That which Nicomachus of Gerasa (about 100 A.D.) presents as early Pythagorean number theory origi-

nated, for the most part, at a much later date, as the theory of
over-perfect and defective numbers (sum of the aliquot parts
$\geqq$ the original number), the friendly numbers (as 220 and 228,
each of which is the sum of the aliquot parts of the other one)
and polygonal numbers (first mentioned by Hypsicles about
180 B.C.), and also the demonstration of cubic numbers as sums
of consecutive odd numbers. The somewhat more recent Neo-
Platonist Theon of Smyrna (about 130 A.D.) found the sum
(easily derivable therefrom) of the first cubic numbers. About
the same time, Sextus Empiricus (about 150 A.D.) presented a
strongly monistic picture of the Pythagorean teachings (of
Poseidonius?—about 70 B.C.), in which the generation of a line
by a moving point, etc., was thoroughly discussed and rejected.

Totally different in nature was the important arithmetical
work of Diophantus of Alexandria (about 250 A.D.), who
adopted and continued the Egyptian-Babylonian tradition in
an un-Greek fashion. He presented, in a form rather markedly
algebraized by the introduction of expedient abbreviations,
interesting examples which show us that he was fully familiar
with the solution of linear equations. When dealing with more
than one unknown, he helped himself by the introduction of
an expediently selected single unknown; when, in cases of
quadratic equations (there are serious gaps in the text here),
there were two positive solutions, only one of them (the greater
one) was accepted. His complete mastery of form becomes evident in his handling of quadratic equations with several unknowns; there are even occasional deviations from the dimen-
sion principle to which the Greeks otherwise adhered so
strictly. The solutions are given in rational, never integral
form. The Moslems became acquainted with Diophantus com-
paratively late, and they continued only his algebraic approaches; it was in the early Baroque that the further development of Diophantine thought produced, first, algebra using
literal notation and, secondly, modern number theory.

The legacy of the great Greek mathematicians was once again

epitomized by the Alexandrian Neo-Platonists (250–650 A.D.). Pappus of Alexandria (about 320) produced a valuable encyclopedic work in his *Collections,* with interesting abstracts from surviving as well as lost writings of Euclid, Archimedes and Apollonius, and also excellent contributions of his own, among which specific mention is due to the propositions concerning projective elements, to the comments on the treatment of extreme values, to the so-called theorem of Guldin on centroids of solids of revolution, and to the researches on the quadratrix, on the spiral of Archimedes and spherical spirals, as well as on helicoidal surfaces. Most of these subjects became points of departure for the investigative methods of the 17th century. The study of Euclid produced the generalization of the theorem of Pythagoras which has been named for Pappus. Of the *Almagest* commentary only parts have survived.

This work was carried on by Theon of Alexandria (about 370), who gave the world also a new edition of Euclid's *Elements,* so sorely needed in those days because of the deterioration of the texts available; the new edition was expanded, of course, by epigone-like insertions of scant importance. Theon's daughter, Hypatia (370?–415), also wrote commentaries on Apollonius and Diophantus, no part of which has survived to our era. Serenus of Antinoia (about 400), whose monograph on the plane sections of the cylinder and of the cone is known to us, was in all likelihood just a little older.

Of the Athenian Neo-Platonists we have to mention Domninus (about 450) and his commentary on Nicomachus, Proclus Diadochus (410–485) and his extensive Euclid commentary, Marinus (about 500) and his introduction to Euclid's *Data,* Simplicius (about 520) who wrote important commentaries on Aristotle, and Damascius (about 520) with an unimportant essay on regular solids which appears in the later editions of Euclid's *Elements* as Book XV. There was also Eutocius of Ascalon (born 480) with his mathematically insignificant but historically informative elucidations of Archimedes and Apollonius, published by Isidorus of Miletus (about 520) who, in collaboration with Anthemius of Tralles (died in 534), built the Hagia Sophia in Constantinople (532–537). Anthemius is given credit for the string construction of the ellipse and for a treatise on the focus of the parabola.

The persecution of pagans in 415, which claimed Hypatia as one of its victims, brought the end of the Alexandrian school of mathematicians; Justinian closed the Academy in Athens (529), but the tradition survived in Constantinople for a long time. In the 9th century, the philosopher Leon encouraged his disciples there to copy mathematical manuscripts. Studious Moslems endeavored about the same time to acquire an encyclopedic knowledge of the entire scientific heritage of the past ages. Thus, despite the unfavorable circumstances, much more original lore was salvaged out of the inner and outer collapse of the ancient world, to survive the great migration of peoples, than seemed to be the case at first. Old and new knowledge flowed to the western world from India, from Syria, from North Africa, producing in the Renaissance a revival of the interest in mathematics, and then, in the Baroque, a complete transformation and reformation of the absorbed lore—the genesis of modern mathematics.

## CHAPTER THREE

# The Middle Ages
## (approximately 500–1400 A.D.)

### 1. INDIA (500–1200)

THE mathematical advance of greatest consequence achieved by the Indians was the invention of the *decimal position system,* known definitely to have been used since the 7th century. It superseded the juxtapositionally used *Kharoshthi* numerals with their vestiges of a quaternary system (inscriptions from 400 B.C.–300 A.D.) as well as the likewise juxtapositional *Brāhmī* numerals (inscriptions from about 250 B.C.) with individual symbols for units, tens and hundreds. The position system originated from the excellently developed technique of calculation on the wooden tablets customarily in use in India, which were sprinkled with sand, so that the symbols traced on them could be altered easily. At first undoubtedly also the Indian counting-board was divided into columns according to the universal custom; it is no longer possible to determine when the symbol for the blank or void column (i.e., the zero) came into use and the lines were dropped (which innovation completed the transition to the position system). We are acquainted with collections of problems and economic texts even from the early period (e.g., the *Bakshālī* manuscript, dating in all likelihood from the 6th century); they embody teachings on the direct, indirect and synthetic reasoning.

Of interest is the accomplished algebraization, in the course of the following centuries, of the calculation process; this change manifested itself, among other features, in the use of the zero and negative numbers, in the introduction of definite designations (some constructed multiplicatively, other additively) for the unknowns and their first powers, in the formulaic treatment of bracketed calculations as well as of the ascertainment of square and cubic numbers, and in the theory of equations. The linear equation was solved at first by means of the *regula falsi*, later directly; in the case of the quadratic equation, treated uniformly as early as the 9th century, the Indians recognized the twofold nature of the solution and did not object to a negative one; they knew also that the square root of a negative number could not be determined. In systems of linear equations, in the 6th century symmetrically solvable types occupied the center of interest; later very many special rules appeared, including also a general one, namely the comparison method for two linear equations with two unknowns. Despite the wealth of data, no uniformity was achieved with respect to quadratic equations with more than one unknown. In that respect, as in the rational approximation of square and cube roots, Babylonian influence must have been a prominent factor, whereas reductions by roots of polynomials must have been influenced by Euclid, but everything was purely arithmetically oriented and quite considerably improved by that.

The divisor problem $x = q_1 t_1 + r_1 = q_2 t_2 + r_2$, came up probably in connection with astronomical questions (the calendar). Āryabhata (born in 476) solved it by means of the continued fraction method; shortcuts and improvements followed later. Brahmagupta (born in 598) carried out the integral solution of the equation $x^2 - py^2 = 1$ on the basis of the auxiliary equations $u^2 - pv^2 = -1, \pm 2, \pm 4$; Bhāskara (1114–1185?) reduced the auxiliary equation $u^2 - pv^2 = q$ by means of an appropriately defined parameter, $t$, to the new auxiliary equation

$$\left(\frac{ut + pv}{q}\right)^2 - p\left(\frac{vt + u}{q}\right)^2 = \frac{t^2 - p}{q} = q'$$

and went on with this reduction until obtaining $-1$, $\pm 2$ or $\pm 4$ on the right (cyclic method). Furthermore, he handled general indeterminate quadratic equations (also simultaneous systems) by technically very skillful special arrangements, and also the way of making a square out of $\frac{ax + b}{c}$. There are numerous evidences to indicate links to Diophantus (about 250), but Diophantus never thought of insisting on integral solutions. With respect to the simplest arithmetical progressions and to geometrical progressions in general, the Indians originally did not go beyond the knowledge of the Babylonians and the Greeks; Halāyudha presented Pascal's triangle in the 10th century, and Nārāyana, the sums of figurate numbers in the 14th century.

We have no knowledge of any particularly outstanding geometrical accomplishments of the Indians. An original feat is Brahmagupta's determination of integral right triangles out of $u, \frac{1}{2}\left(\frac{u^2}{v} - v\right), \frac{1}{2}\left(\frac{u^2}{v} + v\right),$ utilized in the construction of quadrilaterals with mutually perpendicular diagonals. The appended area computation from $\sqrt{(s - a)\,(s - b)\,(s - c)\,(s - d)}$ is correct solely because Brahmagupta limited himself to inscribed quadrilaterals. There is no substantiation given; presumably it was a case of a generalization of the so-called Heron's formula. Of importance is the further development of the Greek chord trigonometry, known to the Indians from translations of Alexandrian astrological-astronomical texts, into a half-chord trigonometry. In the writings of Āryabhata, obscure mnemonic verses contain allusions to the computation of a table of sines of the multiples of $3\frac{3}{4}° = 60° : 16$ by continued angle bisection. Varāhamihira (about 500) already resorted to trigonometric formulas. Manjula (about 930) con-

sidered the functions *sin, cos* and *1 — cos* in all four quadrants.
But these contributions to plane trigonometry were mere
by-products in connection with astronomical problems which
were treated at great length already in the Sūrya Siddhāntā
(4th century?). The Indians at first used a graphic procedure,
which may have originated from an imperfect transplantation
of Greek-Egyptian prototypes, but later they adopted a computational procedure, at which stage even the beginnings of
the law of cosines and the law of sines of spherical trigonometry made their first appearance. The Leibniz series for $\pi/4$
and a procedure for improving convergence occur, as isolated
results, in writings originating from the 15th century (*Karana-
paddhati.*)

Indian mathematics was characterized by the aversion to the
geometrical formulation of algebraic problems, and above all,
by the general arithmetical-algorithmical tendency which suppressed the method of proof, heart and core of Greek science.
The Moslems received from the Indians, through the Persians,
the position value system, algebraic ideas and astronomical-
trigonometrical details, but on the whole obviously mere fragments of Indian knowledge; the magnificent methods and results of the theory of numbers, in particular, remained almost
completely ignored.

## 2. THE MOSLEMS (750–1300)

Syrians, Mesopotamians and Persians were the principal
bearers of the Islamic civilization which reached its first golden
age at the court of Baghdad through the intelligent sponsorship of the first Abbāsid caliphs (from 750). The last Hellenistic adepts of Greek philosophy, natural science and medicine,
who had fled from Byzantine intolerance into the Sassanian
empire, and their Persian pupils brought the necessary data
within the reach of the Moslem scholars, including the excellent translations of the writings of Aristotle and his Neo-Platonist commentators. Then came the great East Arabic philoso-

phers, like Al-Kindi (800?–873/74), Al-Fārābi (870–950/51), Avicenna (980–1037) and Algazel (1059–1111), with their own independent studies. In the astronomical-astrological field, the Moslems were disciples of the Indians (Brahmagupta's *Siddhāntā* was translated to Arabic about 775) and the Greeks (Ptolemy's *Tetrabiblos* was translated about 780, the *Almagest* about 833). The Eastern Arabs took over the place value system from the Indians for practical purposes and tables, but they were satisfied with the—impractical—formal transference of the methods of counting-board computation to written calculation (the arithmetic book of Al-Khwārizmī, about 820); in the actual mathematical writings the numbers were not indicated by figures, but were written out in words.

In the 9th century, the principal writings of the Greek mathematicians were translated into Arabic: Euclid was translated by Al-Hayyay (about 820) and subsequently by many others, Apollonius by Al-Himshi and the Banū Mūsā (about 875), Archimedes and Menelaus by Ishāq ibn Hunein and Tābit ibn Qurra (about 890), and finally Diophantus, Heron, Autolycus, Theodosius and Hypsicles by Qustā ibn Lūqā (about 900). In Al-Khwārizmī's *Collection of Problems for Merchants and Testamentary Executors* many denotations and an algebraizing tendency reveal Indian influence, whereas the geometrical parts show the effects of Heron and Euclid (probably an indirect effect). The *Geometry* of the Banū Mūsā contains already revised parts of Ishaq's translation of Archimedes; Al-Māhānī (about 880) expressed the (geometrically given) cubic problems of Archimedes in equation form, and Tābit (826–901) knew how to enlarge on his sources with a critical attitude, factually correct, and how to develop the accepted material independently (e.g., the law of sines for the right spherical triangle, first rule for the construction of friendly number couples).

Distinctly independent mathematical accomplishments of the Eastern Arabs become evident in the 10th century. In the

domain of algebra, Abū Kāmil (850?–930?) carried on the
work of Al-Khwārizmī and followed (e.g., in the power notation and in the handling of indeterminate linear equations)
the views of the early Indians. Abū'l-Wafā (940–998) tried in
his Diophantus commentary to develop further the Euclidean
theory of irrationals. Al-Karkhī (died in 1029) made use of
Abū Kāmil and Abū'l-Wafā; in the fundamental questions
(rejection of the zero and the negative numbers, power notation) he took the Greek attitude, he treated composite irrationals and the extraction of the roots of algebraic expressions.
Ibn Al-Haitam (965?–1039) arrived, in connection with the
cubature of the paraboloid of revolution, at the formula

$$\sum_{k=1}^{n} k^4 = \frac{n \ (n + 1) \ (2n + 1) \ (3n^2 + 3n - 1)}{2 \cdot 3 \cdot 5}.$$

Al-Khayyāmī (1044?–1123/24) presented a systematic arrangement of the cubic equations and solutions by means of the
conics; he was of the opinion that it was not possible to find an
algebraic solution as well. The Moslems never came to share
the more profound knowledge of the Indians in the domain of
number theory; ever after Abū'l-Wafā, they kept following
more and more strongly in the footsteps of Diophantus. Al-
Khokhendī (died about 1000) believed that he could prove
the impossibility of $x^3 + y^3 = z^3$, and Al-Karkhī treated indeterminate quadratic problems without insisting on integers.

Great progress was achieved in the domain of trigonometry.
Al-Habash (770?–870?) introduced the tangent function and
tabulated *sin, tg, cotg, cosec;* Al-Battāni (850?–929), the astronomer famous for his planetary theory, who treated of the
entire Indian and Hellenistic literature available to him, preferred the sine trigonometry as the more expedient method of
calculation. Abū'l-Wafā improved on the accuracy of the tables
(sexagesimal ones), computing the values correctly to eight
decimals; he made the still cumbersome calculations on the
spherical triangle easier by converting the "rule of the six

quantities" (Menelaus) into a "substitute theorem" ("rule of the four quantities") through the use of a transversal perpendicular to two sides of the triangle. Abū Nasr (about 1000) discovered the sine law, Al-Bīrūnī (973–1048) completed a recalculations of the trigonometric tables applying the Archimedean premise, and At-Tūsī (1201–1274) presented for the first time a self-consistent trigonometry of plane and spherical triangles, making use of the polar triangle. His influence is dominant in the great tables of Ulūkh Beg (1393–1449), which were computed according to the methods of the capable Al-Kāshī and which list the sexagesimally tabulated functions to an accuracy to 17 decimals.

In the field of geometry, we must mention Abū'l-Wafā for his constructions with a fixed opening of the compasses, and Ibn Al-Haitam for the quadrature of the lune, transforming it into a general right triangle, as well as for the problem, from the *Optics*, which calls for the determination on a circle of the points the sum of whose distances from two given points represents an extreme value. A very significant accomplishment was the cubature of the solid generated by the rotation of a halfsegment of the parabola about a line parallel to the tangent at the vertex, while Al-Karābīsī's study of the torus shows amazing formal shortcomings. In conclusion, let us mention At-Tūsī's ingenious comments on the parallel postulate in his edition of Euclid, which is outstanding otherwise as well.

No more than mere fragments of the mathematical knowledge of the Eastern Arabs reached the West where several independent cultural centers developed under the protection of the Omayyad caliphs of Cordoba (736–1031), namely, first, the Indian arithmetic using the counting-board with number forms peculiar to it (the so-called "dust numbers"), later a little trigonometry and astronomy and some of the Arabic translations of classical Greek authors. Al-Zarqāli (1029–1087?) failed to become aware of the application of the tangent function. Jābir ibn Aflāh (died about 1145) even returned to the

Ptolemaic chord trigonometry, but he did treat the spherical triangle really skillfully on the basis of the rule of the six quantities. In the realm of philosophy, mention is due, beside Avempace (1106?–1138) and Abubacer (1100?–1185), chiefly to Averroës (1126–1198) with his carefully commentated edition of Aristotle which exercised such strong influence on Scholasticism in its Golden Age.

Whatever philosophical and mathematical knowledge reached the Christian-Latin West was transmitted to it mostly by Spanish Jews. The translator's school of Toledo, scene of the combined activities (about 1150) of Johannes of Seville, Domenico Gundisalvi and Gerard of Cremona, must be mentioned in particular. Moslem mathematics had very little influence on the Byzantine scholars whose borrowing from their neighbors was practically limited to the Indian figure reckoning (presentations of M. Moschopulos, about 1300, and N. Rhabdas, about 1340). With the fall of the Caliphate of Baghdad (1258) there crumbled also the scientific strength of the Eastern Arabs; the only product of the later period worthy of any mention at all is the encyclopedic work (to be sure, a rather insignificant one) of Behā ed-Din (1547–1622). Our knowledge of Moslem mathematics is incomplete because many a valuable accomplishment has been forgotten and still others have never been revealed to the public.

### 3. The Chinese (200 b.c.–1300 a.d.)

The earliest arithmetic books of the Chinese were collections of problems of the type generally encountered at the start of periods of mathematical development. We are acquainted with the *Chiu-ch'ang Suan-shu* (about 250 b.c.) from a revised edition published about two generations later. It presents, among other details, the treatment of systems of linear equations on the basis of determinant-like prescriptions, also quadratic equations, Pythagorean triangles, and interesting geometrical formulas of approximation, as, for instance, $\frac{h}{2}$ $(s + h)$ for the segment of a circle ($s =$ chord, $h =$ height). The *Wu-ts'ao Suan-king* (1st century A.D.) presented the example

$$x = 3t_1 + 2 = 5t_2 + 3 = 7t_3 + 2\,[=q_i t_i + r_i]$$

which has been cited again and again by the Indians, the Arabs and the western mathematicians. The original Chinese method by means of the "great extension" is based on the solution of the auxiliary equations $q_2 q_3 u_1 = q_1 v_1 + 1$, etc. (in as small numbers as possible, by guessing). Then the remainder obtained in the division of $\sum q_2 q_3 r_1 u_1$ by $q_1 q_2 q_3$ is a solution. A western influence is revealed by the remark of Tsu Ch'ung Chi (430–501) that $\frac{22}{7}$ is an inexact approximation for $\pi$, but $\frac{355}{113}$ is a more exact one; the latter value was derived from $3.1415926 < \pi < 3.1415927$. The interesting cubic problem of *Wang Hs'iao-t'ung* (about 625) calls for the solution of

$$x^2 + y^2 = z^2,\ xy = 706\frac{1}{50},\ z = x + 36\frac{9}{10}$$

It is correctly solved, but no derivation is given.

In the 11th century, printed copies of the arithmetic books of Liu Hui (about 263 A.D.) and Ch'ang K'iu-kien (about 575) became available. Writings dating from the 12th century contain descriptions of the calculating apparatus containing balls, the *suan-pan,* although that excellently designed instrument had probably been invented still earlier; it was, in all likelihood, a product of an adroit development of the older abacus, and in practical application it was equal in value to the figure reckoning of the West. About the middle of the 13th century Ch'in Kiu-shao and Li Yeh (1178–1265) developed a highly effective repetitive method for the numerical solution of equations, to which their contact with the Eastern Arabs may have inspired them. The same Eastern Arabic influence must have been responsible also for the trigonometrical treatment of astronomical problems, perhaps also for the summation of the square numbers by Ch'on Huo (1011–1075) and Yang Hui (about 1260)—the latter treated also $\Sum$ $(\frac{k}{2})$—and for the extensive operation with algebraic transformations which finally culminated in the formulation by Chu Shi-kié (about 1500) of Pascal's triangle. With these remarkable accomplishments, native Chinese mathematics came to an abrupt end, although the Japanese developed it further in an interesting and original fashion in the 17th and 18th centuries.

### 4. THE EARLY CHRISTIAN MIDDLE AGES (200–750)

The faithful of the first Christian congregations were concerned primarily with the propagation of the gospel, with the spreading of the revelations of their faith, with the freedom of worship, and with the moral rebirth of man. This mission (an especially urgent one for them because a mistaken interpretation of passages of the gospels made them expect the end of the world to be forthcoming very soon) made them enemies of "pagan" learning. Tertullian (160?–after 222) still regarded philosophy as the real fountainhead of all heresy and combatted worldly knowledge as folly before God. This attitude underwent, however, a fundamental change as the disappointment of the eschatological expectations was followed by the understanding that the existing world, with all its weaknesses and properties, must be conquered for the gospel. The new aim was to build scientific foundations under the faith, and the maxims of the philosophers were no longer regarded as inimical, but recognized as containing a great measure of truth, although still imperfect in comparison with the revelations of the faith. Origen (185/86–254/55), since 203 a teacher and later (until 231) head of the catechistic school of Alexandria, was the creator of a theology which followed the ideas of Plato and rested consciously on contemporary science. Like Plato, he regarded geometry as the methodical prototype. The same view was advocated in the East by Eusebius of Caesarea (265?–339/40) and Gregory of Nyssa (335?–395?), and in the West by Marius Victorinus (4th century), the capable translator of numerous Neo-Platonistic writings and of a few treatises of Aristotle on logic, and above all by his great pupil, Augustine (354–430), bishop of Hippo. They were all filled with a conviction of the profound significance of mathematical insight and were devoted to mathematical studies, even if they were not active themselves as discoverers, commentators or educators in this domain.

To be considered to possess a well-rounded education, a man had to have completed a curriculum in the seven liberal arts (*artes liberales*) which were divided into two groups: The *quadrivium* which comprised the natural-science subjects, namely arithmetic (some number theory and number symbolism), geometry (geography and natural history), astronomy (determination of the dates of the religious holidays) and music (theory of intervals)—and the *trivium* which consisted of the linguistic subjects, namely grammar, rhetoric (epistolography and the study of law) and dialectic (public speaking and disputation). The first still existing outline of this curriculum is by Martianus Capella (about 450), but it is treated more thoroughly in the monographs of Boethius (480?–524?) written about 500, of which the *Arithmetic* is an abridged version of the *Introductio* of Nicomachus, while the *Astronomy* is based on Ptolemy, and the *Music* on the writings of Euclid, Ptolemy and Nicomachus on this subject. Of a translation of Euclid's *Elements* merely fragments are known. Cassiodorus (480?–575?), a disciple of Boethius, published a new compendium of the *artes liberales* about 554, which remained extremely popular for several centuries because of its lucidity. He may have been also the author of a widely used *Instruction for the Calculation of the Date of Easter* (562). Cassiodorus and Boethius supplied the prototypes for the synopsis pre-

sented in the first three books of the *Origenes* of Isidorus of
Seville (570–636).

All these presentations are of very little value insofar as
their contents go; nevertheless, they did salvage a little number
theory and a few fundamental geometric concepts from the
tempests of the migration of the races. Only the writings of the
Venerable Bede (674–735), a man highly educated for his time,
contain something more, as, for instance, the oldest written
elucidation of the so-called "finger-reckoning," i.e., the
method of indicating numbers by holding the hand and fingers
in certain pre-arranged fashions. This method had been in general use for centuries, but had never before been recorded
systematically. In those troubled times the interest in the
mathematical thought processes was kept awake not by the
sketchy and inadequate teaching of the *artes,* but by the course
in philosophy which was established for the later clerics only.
The students of that course carefully analyzed and studied the
Boethius translations and elucidations of writings of Aristotle,
as well as the commentaries added by some Neo-Platonists, all
of which contained numerous references and comparisons of
a mathematical nature. These were at first merely faithfully
copied by industrious monks and handed on as unfathomed
lore; but they were to play an important part in the revival
of independent mathematical thinking at a later date.

## 5. The Carolingian Pre-Renaissance and its After-Effects (750–1100)

The progressing consolidation of the political conditions
furthered the cause of the sciences and arts, too. Charlemagne
(742–814) acted as an unexcelled patron of culture in 781
when he attached the learned Englishman Alcuin of York
(730?–804) to his court, placed him in charge of his palace
school and of the reorganization of the completely shattered
educational system in the Frankish empire. Alcuin performed
the task with masterly skill. He established a model school in

the Abbey of St. Martin in Tours, entrusted to him in 796. His recapitulatory presentations of the subjects of the *trivium* were written for educational purposes; the *Computation of Easter* was intended to facilitate the execution of the royal decree requiring at least one monk in every monastery to be able to determine the date of Easter without help by anybody else. Alcuin may have been also the author of a witty collection of arithmetic problems which became widely known under the title *Propositiones ad acuendos iuvenes.** Besides Tours, the schools of Ferrières and Corbie acquired an importance in France, as did the schools of Fulda (reformed by Hrabanus Maurus, 784–856), Reichenau and St. Gallen in Germany. The aim pursued consisted at first merely in an improvement of the previous textbooks and of the method of instruction. This purpose was served by the commentaries of Johannes Erigena (810?–877?) and Remigius of Auxerre (841?–908?) on Martianus Capella, and by the German translation, distinctly influenced by them, by Notker Labeo.

The art of elementary reckoning by numbers, hardly even mentioned in the previous theoretically oriented books on arithmetic, was treated thoroughly now. The earliest known elucidation of the calculation on the multi-columned Roman abacus was written by Gerbert of Aurillac (940?–1003; he became Pope Sylvester II in 999) and served as a prototype for a great many similar books on computation. Gerbert operated already with numbered counting beads, with which he had become acquainted while staying in Spain (967–970). The procedure used by him differed from the modern method of reckoning by numbers primarily through the absence of the zero. Gerbert's geometrical knowledge was still very scant and did not exceed basically that of the Roman land-surveyors (the *agrimensores*). Only as abbot of Bobbio did he come across an unexpected treasure in the library of that monastery, namely parts of a translation of Euclid by Boethius. The origi-

* "Problems for the Quickening of the Minds of the Young."

nal of a *Geometry* compiled by Gerbert (possibly in Rheims between 983 and 991) had been lost. We know only later revised versions, but we can reach various inferences about the contents of the original from a discussion, conducted (about 1025) by correspondence among several teachers of mathematics in the Rhenish monastic schools, on the concept of the exterior angle and of the sum of the angles of a triangle. We find, for instance, that the Euclidean teachings on parallel lines were still unknown, nor had anything been learned on this subject from the Aristotle translation and commentary by Boethius, so that the sum of the angles of a triangle had to be determined by purely experimental methods. The low level of geometrical knowledge is demonstrated even more clearly by a scientifically unimportant treatise of Franco of Liège on the squaring of the circle (about 1050).

The interest in mathematical subjects did not diminish by any means among the leading scholars of the Carolingian "renaissance" and of the next two centuries, in comparison to the previous epoch; the level of average learning rose, on the whole, in connection with the improvement of the educational system, but still not to the point where the intellectual heritage would be completely absorbed, let alone developed further by independent research. That point could be reached only after the acquisition of still further knowledge, partly from re-emerging Greek originals, partly from the translations from the Arabic of works of Moslem scholars.

### 6. The First Great Translations and their Effect on Early Scholasticism (1100–1200)

Beginning in the earliest years of the 12th century, the scholars of the Latin West were benefited by an ever more abundant afflux of new knowledge, on the one hand from the translations of Greek originals (Southern Italy, Sicily, Bzyantium), some just re-discovered, others just made accessible again, and on the other hand from the translation and adapta-

tion of Arabic texts (Spain). At the same time, under the influence of Anselm of Canterbury (1033–1109) and chiefly of
Peter Abelard (1079–1142), there developed a clear technique,
based on simple rules, of theologico-philosophical inquiry, the
so-called *Scholastic method*. This method can be summed as
follows: One lists first the views (regarded fundamentally as
binding) of the Fathers of the Church and other authorities
on a given subject of discussion, weighs carefully the arguments
for and against the individual views, reaches the decision on a
purely ratiocinatory basis, and tries to reconcile and eliminate
any possible dissenting opinion by the choice of a sufficiently
general and all-embracing point of view. This technique was
developed by Abelard on the ground of a then widely used
grammar of Priscian (about 550) in the sense of a purely
formal logic; later (roughly from 1175 on) it was combined
with Aristotelian logic into the so-called *logica moderna* which
was permeated thoroughly by general mathematical conceptions.

Of the logical treatises of Aristotle originally only the *Prae-
dicamenta* and *Peri Hermeneias* were known in the translation by Boethius (the so-called *vetus logica*); the *Analytica
Priora* and *Posteriora*, the *Topica* and the *Sophistici Elenchi*
(the so-called *nova logica*) were made part of the educational
program by Thierry of Chartres (died about 1150) between
1136 and 1141, but not in the translation by Boethius (which
was already very rare at that time, and is completely unavail-
able today) but in the translation completed by Jacob of Ve-
netia in 1128. Like a great many other masters of the school
of Chartres, Thierry was an enthusiastic admirer of the ancients. His extensive textbook, the *Heptateuchon*, which remained forever in the manuscript state, contained Latin excerpts from forty-five of the best authors of Antiquity. He used,
among other books, the *Planisphaerium* of Ptolemy and the
arithmetic book of Al-Khwārizmī in the translation made
from the Arabic by Hermann the Carinthian (in 1143), also

an approximately contemporary translation, erroneously attributed to Boethius, of Euclid's *Elements* (without proofs). Even greater was the influence of Gilbert de la Porrée (1080?–1154), the most important logician of the school of Chartres, whose much commentated *Sex Principia* remained in use at the University of Paris from the 13th century on as the authoritative supplement to the reading of Aristotle's treatises on the subject.

A school of translators was formed in Toledo about the same time under the energetic sponsorship of the Archbishop Raymund (1126–1151); Plato of Tivoli, Rudolph of Bruges, Daniel of Morley and Robert of Chester were its most prominent members. Knowledge was disseminated mostly by Jewish scholars, for instance by Abraham bar Hiyya (1070?–1136?), who was the author of an introduction to geometry based on the writings of the *agrimensores,* Heron and Euclid. Johannes of Seville, a baptized Jew (died about 1153), translated from the Arabic and Hebrew word for word into Castilian, and his collaborator, Domenico Gundisalvi, re-translated it word for word into Latin. Logical and philosophical treatises of the Arabs and Greeks were given preference, but also astrologico-astronomical, medical and mathematical writings were made accessible. No sooner had these works become known than they were received by the Latin-speaking scholars with avid interest, despite the serious shortcomings of translation.

Of a much higher caliber was the ability as a translator of Adelard of Bath (1075?–1160?), a man who travelled far and wide, knew well Arabic and Greek and was an outstanding expert on Aristotle. He translated the arithmetic book of Al-Khwārizmī and Euclid's *Elements* from the Arabic, Ptolemy's *Almagest* from the Greek (between 1153 and 1160). He emulated a Byzantine codex which had been brought to the Norman court of Palermo by Henricus Aristippus. About the same time, the Sicilian admiral Eugenius translated Ptolemy's *Op-*

*tics* from the Arabic. Another translation of the *Almagest* was made (1175) by Gerard of Cremona (1114–1187) through whose efforts the West gained access not only to astronomical treatises, but also to portions of the natural-philosophical writings of Aristotle and the Arabic commentators, as well as to the valuable Euclid commentary of An-Nairīzī, with its numerous references to the views and findings of earlier Greek and Moslem mathematicians.

Hugo of St. Victor (1096–1141), one of the most influential of the *masters* of Paris, had already classified mathematics in his general theory of science (the *Didascalion*) as a purely theoretical discipline in the same category as theology and physics. His arguments were as follows: The subject of mathematics is quantity abstracted from reality, its task is to unravel complicated reality by means of logical ratiocination; it relates to the external form of things, namely: arithmetic to numbers, music to proportions, geometry to extent, and astronomy to motion. Unshakable truths, he reasons, can be won only through the application of reason; experience is subject fundamentally to the possibility of delusion. His pupil, Clarenbald of Arras (about 1160), dispensed already with a sharply delineated metaphysics and replaced it by the *speculatio mathematica* and *theologica*. Alanus ab Insulis (1120–1203?), the famous *doctor universalis*, attempted in his *Regulae Theologiae* to present an axiomatically constructed picture of theology on the basis of the mathematical-deductive methods. Nicholas of Amiens, in his *Ars Catholicae Fidei* (about 1190), expounded a rational theology which he linked closely to Euclid's *Elements*, with complete indifference for the authority of the Fathers of the Church and for any reconciliation of different doctrinal attitudes by the dialectic methods. This indicates with what impact the general interest in the mathematical-logical method extended into the innermost areas of theology. But that interest did not spring from deeper knowledge of the details of mathematics; it followed rather from the accepted judgments of values expressed by Boethius, the Neo-Platonists and the Aristotle commentators. The factual knowledge comprised hardly more than the contents of Euclid's books on plane geometry and the Indian number reckoning and its application to the solution of simple equations in the texts. The most important parts of all this were destined to belong soon to the standard stock of higher education.

## 7. THE GOLDEN AGE OF SCHOLASTICISM
### (13TH CENTURY)

A great ideological conflict ensued in the 13th century: The teachings and views inherited from the Neo-Platonists and from Augustine vied with Aristotelianism which had been embraced by leading Moslem and Jewish scholars, such as Avicenna (980–1037), Avicebron (1020/21–1069/70), Averroës (1126–1198) and Moses Maimonides (1135–1204). Domenico Gundisalvi had advocated an expansion of the previous program of studies and the inclusion of the most important Aristotelian writings on the curriculum of the *artes liberales* as early as 1150. The second half of the 12th century saw the great cathedral and monastery schools gain ever more and more in importance and become independent as "Universities."

About 1200, the school of the Cathedral of Notre Dame and the St. Denis, St. Géneviève and St. Victor convent schools were combined in the University of Paris which from then on comprised four faculties (*artistae, theologi, decretistae* and *medici*). The Universities of Oxford and Cambridge, organized along similar lines, came into existence almost at the same time. The pupils entered the faculty of *artistae* approximately at the age of ten and spent six years there. Only after passing a final examination could a pupil be admitted to one of the other faculties. The theological faculty was the most influential and most respected one; its curriculum covered eight years and ended with a ceremonious examination and disputation.

A graduate who wished to acquire higher honors was given first, as a *baccalaureus*, a subordinate teaching position, then if he proved successful, after a reasonable period of time, the *magisterium* (the authority to teach in a certain faculty) and *doctorate* (the authority to teach in every University). Mathematics was a specialty of its own within the faculty of *artistae* and extended over subjects of the *quadrivium*, but it was becoming more and more overshadowed in importance by logic, philosophy and theology.

The intellectual leadership in the Universities soon passed into the hands of the two mendicant orders, the Franciscans (founded in 1210 by St. Francis of Assissi, 1182–1226) and the Dominicans (founded in 1215 by St. Dominic, 1170–1221); their membership included the most important students of the century. In Oxford, Robert Bacon donned the habit of the Dominicans, in 1225, Adam of Marsh (died in 1258) that of the Franciscans, in 1226; in Paris, the Dominicans won the first teaching pulpit with the promotion of Roland of Cremona in 1229, and a second one came under their control when Johannes of St. Aegidio joined them in 1230, while the Franciscans won a *cathedra* by gaining Alexander of Hales (1175?–1245) as a member. The Orders offered to their future Fathers, moreover, also the facilities of their *studium generale*, and the *provinciale*, for the perfection of their learning.

From the very beginning, a heated scientific rivalry existed between the teachers in religious Orders and the members of the secular clergy. In Oxford, the mathematical, natural-scientific and linguistic disciplines were cultivated in particular, Paris laid the emphasis on philosophy and theology. In the first half of the century, all faculties tried to adopt and adapt the new Aristotelian, the relatively pseudo-Aristotelian teachings and to combine them with the Augustinian tradition; as a result of the arising conflicts, the Aristotelian philosophy of nature was repeatedly outlawed in Paris (1210, 1215, 1231). Nevertheless, the forbidden writings were carefully studied and commentated by the most respected teachers from 1231 on, and in 1255 they were officially included in the curriculum of the *artes,* because people had learned in the meantime how to separate the genuine Aristotelian writings from the imitations, which were the only ones found objectionable.

As long as the Augustinian views remained predominant, mathematics also continued to occupy a center of general interest. William of Auvergne (a Paris master, died in 1249) was a representative of this view; he declared that the law of contradiction, the law of the excluded middle and the axiom of the whole and the part were the supreme principles of all science. Robert Grosseteste (a master in Oxford, 1175–1253) translated Aristotelian and Neo-Platonic writings from the Greek; he was a very independent thinker and regarded the clerical and scientific authorities with strong reservations. He looked at mathematics as the indispensable prerequisite to the philosophy of nature. Most interesting was his metaphysics of light which was influenced by Neo-Platonistic ideas and had points in common with similar speculations of William of Auvergne and Alexander of Hales. Of novel character was his attempt to reach a mathematical-dynamic foundation by the application of the principle of the greatest effect, in which he anticipated quite modern relativistic ideas. Robert's pupil, the Franciscan Roger Bacon (an Oxford master, 1210?–1295?), tried to present a comprehensive picture of all secular science. He was familiar with Euclid's *Elements,* the *Optics* of Euclid as well as of Ptolemy, the latter's *Almagest* and the *Sphaera* of Theodosius, furthermore the *Treatise on Isoperimetrics* and individual portions of the writings of Hipparchus, Apollonius and Archimedes; and he wrote an interesting book on Optics (*"Perspective"*) which his mathematically substantially less gifted pupil, Johannes Peckham (master in Paris and Oxford, 1240?–1292), condensed into a widely used, but insignificant abridged version. The Franciscan Roger of Marston (master in Oxford, about 1280) still professed, in agreement with

Averroës, that mathematical insight was deeper than the insight in the natural sciences, not with respect to the clarity of knowledge but almost entirely so with respect to the status of knowledge.

The second half of the century brought the decisive turn toward Aristotelianism among the Dominicans. Albertus Magnus (1208?–1280) conceived the great plan of making the entire available knowledge of the Greeks, Arabs and Jews accessible, and to present it critically, to his contemporaries. In matters of the faith he had more confidence in Augustine than in Aristotle, but in natural science Aristotle was his supreme authority. Albert, like Roger Bacon, emphasized the necessity and significance of the natural-scientific experiment. His observations of the skies made him reject the Aristotelian cosmology and espouse the Ptolemaic system. Albert gained fame also as a mathematician; in all probability, he was the author of a revised version of the Euclid commentaries by An-Nairīzī which contained new material (primarily Arab teachings).

Also Albert's great disciple, Thomas Aquinas (1225/26–1274), spoke of Augustine strictly with the greatest reverence, but in the philosophical substantiation of the Christian doctrines—a matter particularly close to his heart—he deviated from the Neo-Platonist course and let himself be guided entirely by Aristotle. Together with his friend, the Franciscan Bonaventura (1221–1274), he overcame the opposition of the secular clergy at the Paris University to the mendicant orders (1252–57). He combatted passionately the Latin Averroism of the faculty of the arts (principal representative: Siger of Brabant, 1235?–1282), according to which the Aristotelian doctrinal views were regarded, exaggeratedly, as the supreme authority in matters of religious truths, too. During his second *magisterium* in Paris (1269–1272) he got into an open fight with the Franciscans and the older Dominicans who worked toward a reconciliation of the Aristotelian system with Augustine; a public clash with Johannes Peckham (1270) was followed by the suppression by the Church of Latin Averroism (1270, intensified in 1277) and in conjunction with that, under a misinterpretation of the intentions of Thomas Aquinas, by the temporary condemnation of individual Thomistic teachings (1277, 1284, 1286).

The aims pursued by Thomas included a reliably guaranteed text of the Aristotelian writings, and the elimination of apocryphal insertions. This purpose was served by a revision; in respect to this a re-translation from the Greek was written (1260–70) by the Dominican William of Moerbeke (1215?–1286). The strength of the Golden Age of Scholasticism was spent in the subsequent controversy about Thomism, which resulted in ever more violent conflicts between the Franciscans who remained loyal to Augustine and their adversaries, the Dominicans, Hermits of St. Augustine, Cistercians and Carmelites; secular science was not furthered at all by these discussions.

Mathematical instruction in the faculty of arts was limited to elementary calculation and some geometry and astronomy. Among the widely used presentations deserving of mention are the meager arithmetic book of Johannes of Sacrobosco (Paris, 1200?–1256?) and the didactic poem of Alexander of Villedieu (about 1225), also the *Sphaera* of Johannes which followed the Ptolemaic view, and a *Computus Ecclesiasticus* (1232). The *sphaerica* was opposed by Michael Scot (1175?–1234?), who had translated (in 1217) the *Planetary Theory* of the Western Arab Alpetragius, a text written about 1200 and setting forth the old Aristotelian teaching of the concentric celestial spheres. The ensuing discussion was decided, on the basis of the material obtained by observations, very rapidly in favor of the epicycles of Ptolemy.

Michael served as astrologer to Frederick II (1194–1250, Roman Emperor since 1215), a beneficial patron of secular sciences. The first professional mathematician of the West, Leonardo Fibonacci of Pisa (1180?–1250?), who rose from the merchant class and had not studied at any University, dedicated his valuable and informative treatises to this Emperor and to his sponsor, Michael, namely: The arithmetic book *Liber Abaci* in 1202 (second edition in 1228), the *Practice of Geometry* (1220), the unfinished *Book of the Squares* (1225), and an anthology of problems (*flos*); he engaged in disputations on the subject matter of the theory of numbers contained in the last two writings, in the presence of the Emperor, who was kindly disposed to him. Leonardo drew his support as much from Arabic as from Greek sources. For example, in geometrical topics he drew upon Banu Musa, in algebra and number theory upon Diophantine tradition. Nevertheless, in all matters of selection, organization, procedure of proof and alteration in the statement of problems, he remained thoroughly independent.

Jordanus Nemorarius—identified in all likelihood mistakenly with Jordanus Saxo (died in 1237), second General of the Dominicans—seems to have been a little younger than Leonardo. He was the author of several algorithmic treatises in which, possibly under the influence of the Aristotelian system of notation, he replaced certain numbers arbitrarily by letters. But he did not go beyond the imperfect experimental stage, on the one hand, because there were no symbols for every operation, and on the other hand, because, in the course of a computation, the same quantity would be denoted, not by consistent use of the same letter, but by different letters. Here, as in his original *Geometry*, Jordanus used Moslem sources. He may have known Arabic. That would speak for the fact that he is credited with the authorship of the Latin translations, based on Arabic versions, of Ptolemy's *Planisphaerium* and of the *Treatise on Isoperimetrics* by Zenodorus.

We owe to Johannes Campanus of Novara (about 1260) a contemporary of Jordanus, the translation, from the Arabic, of Euclid's *Elements* which remained the authoritative version for more than two centuries. The text appears to have been treated with care and ability, with additions by the editor here and there, influenced mostly by the nature of the original text. The question of whether or not the angle of contingence (between the arc and tangent of the circle) has the attribute of magnitude was thoroughly investigated. Campanus, associating the concept of an area with the angle, affirmed the attribute of magnitude; to him the ratio of the angle of contingence to the right angle was in the nature of an irrational, although of a different kind of irrationality than that between the side and the diagonal of a square. On the other hand, he denied the general validity of the theorem of mean value (that which can become greater and smaller, must become equal, too). Later, problems of this kind were taken up with the greatest interest from a philosophical point of view. Campanus also wrote commentaries on the *Sphaerica* by Menalaus and that of Theodosius, and probably also on Jordanus' translation of the *Planisphaerium* by Ptolemy. The *squaring of the circle* ascribed to him—an inferior product of little value, which arrives at the result $\pi = 3\frac{1}{7}$—was certainly not his handiwork, but presumably Johannes Peckham's.

At the same time that Campanus was working on his Euclid translation, William of Moerbeke (1215?–1286) was preparing his great translations of Aristotle and Proclus from the Greek for Thomas Aquinas. Between 1268 and 1272, he mustered the courage the tackle even writings of Heron, Ptolemy and Archimedes. His almost literal translation—by no means surprising in view of the difficulty of the subject—had no effect whatsoever at first. Translations of unknown authorship of other small parts out of the Greek Aristotle commentaries—for instance, an abridged presentation of the squaring of the lune by Hippocrates, Plato's interpolation of two geometric means, the *Treatise on Isoperimetrics* by Zenodorus, the *Squaring of the Circle* by Archimedes and parts of his treatise on Spirals—were in circulation since the end of the 13th century, and occasionally, they would be quoted, too. One of the friends of William of Moerbeke, Witelo, was the author of a book on *Optics,* taken for the most part word for word from Alhazen (Ibn Al-Haitam) and containing also passages from Euclid, Apollonius and Ptolemy. The best accomplishment in the realm of optics in this period was that of Dietrich von Freiberg (1250?–1310?), a Dominican and follower of Augustine: the absolutely correct explanation of the rainbow.

Ramon Lull (1235?–1315), implacable foe of Latin Averroism, stood absolutely remote from this development. He believed that he could achieve an all-embracing scientific system, the *scientia generalis,* through his *Ars Magna,* based on a methodical combination of concepts (the essential portions of which were created between 1265 and 1274). This was the first, even though still imperfect, attempt at the construction of a general axiomatic system for all science. He remained totally incomprehensible to his contemporaries, but attracted increasing attention after the earliest years of the 16th century, and above all he had a decisive influence on Leibniz.

Scholasticism in its so-called Golden Age, dominated by the philosophical conflict between Neo-Platonism and Aristotelianism, had little interest in technical mathematical studies which met with well disposed attention only on the fringes; the level of knowledge rose above all in the field of geometry from the time when parts of the Euclid translation by Campanus were included in the mathematical curriculum of the faculty of *artistae.* The first signs of the approaching intensified cultivation of mathematics as a discipline *per se* are already recognizable in this period of history.

## 8. LATE SCHOLASTICISM (14TH CENTURY)

Scholasticism exhausted its creative power in the next century. The futile struggle between the Thomists and their opponents, who were guided henceforth by the critical activities of the Franciscan John Duns Scotus (1270?–1308; repeated attempt at a reconciliation of Augustine and Aristotle), continued with unabated acuity. Those who did not wish to participate in these controversies sided with the *via moderna* (a modified nominalism) of William of Ockham (1300?–1349/50), whose scientific achievements lay primarily in the realm of logic. The new outlook flourished (despite prohibitions by the Church in 1339–40) in the schools of superior education in Oxford, Paris, Vienna (founded in 1365) and Erfurt (founded in 1392); its exponents were interested very strongly in mathematics and the natural sciences, and they paved the way for new concepts in these fields.

Let it be mentioned quite cursorily that the arithmetic book of Johannes de Sacrobosco was still very popular and was improved in many a respect, formally as well as in contents, by the commentary of Petrus de Dacia (1291); other arithmetic books were published by Johannes de Muris (1290?–1360?), by Levi ben Gerson, a Jew living in southern France (1288–1344), by Johannes de Lineriis (1300?–1350?), and by the bilingual Bernard Barlaam (1290–1348). Some of these presentations contain also the algebraic procedure of the Moslems for the treatment of simple equations. The trigonometric knowledge of Levi ben Gerson, of the Oxford master Richard of Wallingford (1292?–1336), John Mauduith (about 1320), Simon Bredon (about 1380), and of the Paris master Domenicus de Clavasio (about 1370), was drawn solely from Western Arab writings.

New paths were taken by the Oxford master Thomas Bradwardine (1290?–1349), who had the entire philosophical and theological knowledge of his age at his command. His *Arithmetic* is a number theory based on Boethius, his *Geometry* a very concise survey, intended as an introduction for philosophers, of the most interesting features of Euclid's works, based on Campanus, with continual references to Aristotle and Boethius. We find here also a reference to the *Quadrature of the Circle* of Archimedes, as well as an excerpt from the

Treatise on Isoperimetrics, by Zenodorus, and also (in connection with Campanus) an exposition of the question of the angle of contingence and of the incommensurability between the side and the diagonal in the square. The determination of the sum of the angles of a star polygon and a study on filling up space by means of regular solids (with reference to Averroës) go further than Campanus. The treatise, *De continuo,* which introduced a series of similar (still unpublished) essays, was closely linked with these discussions. Thomas Bradwardine treated of problems of continuity and of the possible ways of approaching infinity, of later (cf. Albert of Saxony) questions of extreme values, and of other similar problems. The book on *Proportions* (1328) presents Euclid's theory of ratios and geometric means verbally; in this context, there appears not only (as in the archetype) $a^2:b^2$ as the "double" ratio of $a:b$, but also $\sqrt{a} : \sqrt{b}$, as the "half" ratio. The Aristotelian law of motion, $v = K:R$ (where $v$ = velocity, $K$ = motive force, $R$ = impeding resistance) was transformed—to express Bradwardine's idea in a completely modern fashion—into $v = \log K/R$ and was used in computations, but not checked experimentally. This treatise had a strong influence on Albert of Saxony.

Nicole Oresme (1323?–1382) was a little younger than Bradwardine. From 1356 on, he was a master at the Paris college of Navarre. In his *Algorismus Proportionum* he carried on the reasonings of Bradwardine and developed the principal theorems of computation with fractional powers. This work did not get into print later, but the most essential features of its contents found their way into the publications on the subject by H. de Hangest (died in 1538) and A. Thomas (about 1510). Of still greater importance was the treatise, *De Uniformitate et Difformitate Intensionum,* of which merely excerpts are known today; it was written prior to 1371, and was published later in print in an abridged version prepared by one of his pupils. He expressed in it the increase and decrease of the Aristotelian species of quantities purely symbolically by figures

between straight lines and circular arcs, and the primary interest is not the function and its change, but the figure. In conjunction with the transformation then in progress of the Aristotelian theorem on the motion of projectiles into the so-called
*impetus* theory, this method was applied to develop the theory
of uniformly accelerated motion. Oresme's new theory made
its way into the University curriculum under the designation
*latitudines formarum;* in Cologne, for instance, it was required
knowledge for passing the examination for the Bachelor's degree from 1398 on. Among similar presentations mention is
due to those of Walter Burleigh (about 1340), Richard Swines-
head (about 1350), Marsilius of Inghen (1330?–1396) and
Albert of Saxony (1320?–1390), and to a commentary of Biagio
of Parma (died in 1416). The quantitative computation of the
relationships indicated geometrically in these treatises appears
in Richard Swinehead's *Calculationes,* prototype for many
similarly designed collections of problems. At this time even
infinite geometric progressions were handled and combined
with arithmetic progresisons; the infinite harmonic progression was also making its appearance. Some of all this was recorded in print by A. Thomas. Related questions appeared in
the relevant literature on *Sophismata* and *Insolubilia* in the
writings of Albert of Saxony, Walter Burleigh and William
Heytesbury (1310?–1380).

In view of the fragmentary state of known data concerning
Late Scholasticism, nothing certain can be stated with respect
to Oresme's sources, but it may well be assumed, with prob-
able justification, that he was influenced indirectly by Arab
and Greek prototypes. The fact that an amazingly great wealth
of information was accessible to this great scholar is evident from a note in the French commentary on the pseudo-
Aristotelian treatise *De Caelo et Mundo;* it refers to the treatise *On Spirals* of Archimedes.

According to their intellectual attitude, Bradwardine and
Oresme still belonged entirely to Scholasticism, but already to

that late stage of development which prepared the transition to a new direction of thought. The War of Succession between France and England (1339–1453) emptied the class-rooms and lecture-halls, and prevented the ideas of the last two mathematicians of Scholasticism from meeting with the proper attention and response. The ranks of their pupils did not include one single mathematically creative individual; all of them were capable logicians and indefatigable commentators—as, for instance, Albert of Saxony, whose *Treatise on the Quadrature of the Circle* is a model of the diligently elaborated, carefully pondered composition—but they lacked the ability to produce something new independently. Scientific tradition broke off therefore suddenly and almost abruptly with the end of the era of Late Scholasticism: In mathematical thought, the Modern Age did not follow in the footsteps of the preceding epoch with conscious deliberation, but merely intuitively; it drew its great impetus from the renewed, direct contact with antiquity.

## CHAPTER FOUR

# Humanism
## (approximately 1300–1580)

### 1. THE TRANSITION FROM THE MIDDLE AGES
### TO MODERN TIMES (ABOUT 1300–1500)

IN the Western world, the emergence from the Middle Ages took place in the form of a profound ideological re-orientation; it was no sudden occurrence, but a progressive transformation that went on for several generations. The first signs of a new outlook made their appearance already in the *relievo* work of Niccolò Pisano (1216?–1278) which shows an influence of classical antiquity, and in Giotto's enthusiastically received and developed naturalistic style of painting. The typical yielded its place to the individual, allegoric symbolism to a realistic reproduction of factual circumstances; the individual began to perceive himself as a unique personality and strove for freedom of thought, of expression, of faith. Medieval theocracy, in which the Pope was regarded as the Vicar of God on the Earth, was reduced to mere fiction after the removal of the Papal See to Avignon (1309), and notably as a result of the great schism (1378–1417); the medieval Empire met its end with the tragic downfall of the Hohenstaufens. No longer were Pope and Emperor the dominant factors in the shaping of the social and political structure of the West; that decisive role was taken over by the hereditary monarchies which stood on national foundations and kept gaining in strength continually, and whose intricate plays for power resulted in the development of a diplomacy and statesmanship, dominated later on to a great extent by the theories of Niccolò Machiavelli (1469–1527) and by the power principle.

Latin, heretofore the liturgical language of the Western Church and hence the unchallenged and generally accepted universal language of the West, was gradually displaced by the national tongues and survived eventually solely as a language of science and learning. Authority and tradition lost their former weight and, in consequence of the abuses which had grown to enormous proportions in the legal, social and ecclesiastic domains, they came to be regarded as intolerable shackles.

The systems of medieval philosophy became petrified and, explored as they had been to the last and least detail, they no longer held any attraction; a new intellectual life was developing, not within the pale of the scholastically oriented Universities where traditional learning managed to survive for a long time yet, but in the courts of the ever more powerful princely dynasties, and nowhere as strongly as in Italy. Italy was the fountainhead not only of the artistic, but also of the literary Renaissance which sought and found its model in classical antiquity, in the previously all but totally unknown masterpieces of the great Greek and Roman poets, orators, historians and philosophers, whose writings the scholars now would study until they eventually really understood, and even learned to speak to some extent, classical Latin (how very different from Vulgar Latin and from the strongly deteriorated Church Latin!) and classical Greek.

This literary movement began with Petrarch (1304–1374) and Boccaccio (1313–1375), who attracted Leontius Pilatus, a bilingual but insufficiently trained Calabrian, to Florence as a teacher of Greek (1359–62, translation in prose of the *Iliad*). It received a new powerful impulse in connection with the attempts to bring about a union of the Byzantine Patriarchate and the Roman Papacy. The Byzantine Chrysoloras (1350?–1415), a disciple of the Neo-Platonist Pletho (1355?–1450), arrived in Italy on a diplomatic mission in 1391; from 1397 on, he was active as a public teacher of Greek language and literature in Florence, later in Pavia, Venice and Rome. He completed an important translation of Plato's *Republic* and introduced a large number of younger scholars of great linguistic talent, to his native language, among them Niccolò de Niccoli, a collector of manuscripts (1363–1437, Florence), the historian L. Bruni (1369–1444), translator of portions of the writings of Aristotle and Plato, the diplomat G. Fr. Poggio (1380–1459). His indirect disciples included Francesco Barbaro (1398–1454) and his nephew, Ermolao (1454–1493), translator of Aristotle, in addition, Fr. Filelfo, another translator of Aristotle (1398–1481), possibly also Vittorino da Feltre (1368–1446) and P. C. Decembrio (1399–1477) who translated Plato's *Republic*. The Venetian collector of manuscripts G. Aurispa (1369?–1460) studied in Constantinople itself; his disciple, the poet L. Valla (1407?–1457), was a teacher of Greek in Rome from 1450 on.

The strongest impetus was supplied by the philosopher Pletho, who visited Florence with Johannes VIII Palaeologus (1391–1448; Emperor since 1425) in 1438, to attend the negotiations for the Union, and was the guest of Cosimo I de' Medici (1389–1464; ruler of the republic since 1434) during his stay in Italy. He thought that he would be able to bring new life to Neo-Platonism—to the educated classes a substitute for the Christian ideology and way of life. This purpose was served by the *Comparison of Aristotle and Plato* (about 1440). To Pletho's influence must be attributed the establishment of the *Marciana* (1441), the famous Florentine library, organized according to plans of the highly educated T. Parentucelli (1397–1455; Pope Nicholas V since 1447). A pupil and friend of Pletho was the noble and kind B. Bessarion (1403–1472;

archbishop of Nicaea from 1436), a man of rare knowledge who as advisor of the Byzantine Emperor brought about the (ephemeral) Union of the Churches in Florence (1439), joined the Roman Church (became a Cardinal in 1440), and as a skillful diplomat played an important part in political life; he was the intellectual center of the new endeavors in Italy. He extended benevolent assistance to the Macedonian Theodore of Gaza (1400?–1473?) who fled to Italy before the Turks in 1440, to the grammarian K. Laskaris (1434–1501; in Italy from 1453), and to the philosopher J. Argyropoulos (1416?–1486; in Italy from 1434 to 1441 and then from 1453 on) whose lectures on Aristotle in Florence and then in Rome drew large audiences. His pupils included A. Poliziano (1454–1494) and Johann Reuchlin (1455–1522), the spokesman of the German humanists.

The further controversies about Aristotle within the Mendicant Orders had their counterparts in Byzantium where the older, still purely Aristotelian school was struggling with the Pletho's Neo-Platonically oriented followers. The dispute was continued, in a tempered form, by the Greek immigrants on Italian soil. The ranks of convinced Aristotelians, to whom we owe important new translations of the original writings and of the great Greek commentaries, as well as the first Greek editions, include L. Bruni, Georgius of Trapezunt, Theodore of Gaza and J. Argyropoulos; the Platonists gathered about M. Chrysoloras, L. Valla and Bessarion (the latter was a worker for compromise and conciliation).

Good translations and editions of Platonic writings were provided by the Platonist Academy (from 1450) which enjoyed strong patronage under Cosimo de' Medici and his son, Lorenzo the Magnificent (1448–1492; ruler of the Republic from 1469). Marsilio Ficino (1433–1499), Giovanni Pico della Mirandola (1463–1494) and A. J. Laskaris (1446–1535), the manuscript collector working for the Medicis, are to be regarded as its chief representatives. The Academy directed its efforts above all against the Averroistically oriented Aristotelians of Padua who in 1472 procured an edition of Averroës, on the basis of the Latin translations from the Arabic; an improved edition was produced by the excellently educated A. Nifo (1473–1546) in 1495–97. The strongly naturalistic-nominalistic Alexandrine school broke off from the Averroists somewhat later; the Alexandrines were followers of the views of Alexander of Aphrodisias, the important commentator on Aristotle (about 200 A.D.). Alexandrinism was based on the texts appearing in Volume 3 of the Aristotle edition of 1495–98. Its chief representatives were Pietro Pomponazzi (1462–1524) and the Spaniard Sepulveda (1500–1572?) who provided the translation of commentaries on Aristotle's *Metaphysics* (1527). The strength of the Platonist Academy was broken when Girolamo Savonarola (1452–1498; dictator of Florence from 1494) drove out the Medicis; also the North-Italian school of Averroists and Alexandrines was repressed by the prohibition issued by the Church in 1512. The strife among the different schools of thought in Scholasticism finally faded into the background behind the new natural-philosophical and natural-scientific trend which set its sights on the greatest possible freedom from prejudgment in its careful testing of phenomena on the basis of observation and experimentation, and on the inquiry into relationships determined by natural laws. Within this shift of the point of view, mathematics gradually divorced itself from philosophy and allied itself with the natural sciences; the direction of the activities of the students of nature were guided by mathematical thinking and mathematical cognition from then on.

This process of transformation was accelerated considerably by the invention of the printer's art. Prints from wood-blocks can be traced back to 1423; thirteen years later, in Strassburg, Gutenberg (1400?–1468) signed his first contract "for the practice of an unknown art," meaning the art of printing with movable type. His journeyman, J. Mentelin (1420?–1478), was active as a printer in Strassburg from 1450. In 1452–53, the first bigger print jobs by Gutenberg and his collaborator P. Schöffer (1425?–1502–03) were produced in Mainz. In 1456, P. Castaldi made his first experiments in printing in Feltre. The prosperous printing shop of Schöffer was destroyed when his city fell to Count Adolf of Nassau; a great number of journeymen were scattered to all the winds and they carried the new art over the whole of Europe. The *Bible* was the first great literary work published in print; it was followed by scientific treatises, and from 1474 on (from the Nuremberg printing shop of Regiomontanus), by detailed calendars, astronomical and astrological writings, and from 1476 on also by Greek texts. The first printers were highly educated men, such as Aldus Manutius (1448–1515; in Venice from 1485), the Giunti family in Florence, Venice and Lyons, J. Froben (1460?–1527; in Basle from 1491), J. Bade (1462–1535; in Paris about 1500), and R. Estienne (1503–1559; in Paris from 1526).

Princes who understood art and were friendly toward science soon began to collect books and established sizable libraries at their own expense. The first of these libraries was the *Laurentiana* in Florence (1444), followed by the *Vaticana* in Rome (created by Nicholas V /1447–1455/ and considerably enlarged by Sixtus IV /1471–1484/) and the *Corvina* in the city known today as Bratislava (established by King Matthias Corvinus of Hungary in 1467 when Bratislava belonged to Hungary and was called Pozsony; its collection was scattered to the winds later). With the spreading of the printer's art, and the organization of public libraries, working conditions and the possibility of mutual understanding grew more favorable for scientists than had been the case previously. The individual could enlarge his sphere of action; new ideas quickly found a path to broader publicity. These encouraging circumstances had a substantial part in the advance of mathematics in the Baroque period.

## 2. Renaissance Mathematics (about 1400–1540)

In the domain of mathematics, the change to the Modern Era manifested itself above all in the strong consideration given to purely practical points of view. The center of interest for the historian of science is no longer the learned monk or the scholar teaching in a University, but the reckon master in one of the many commercial cities of Northern Italy, Southern Germany or France, who allied himself with his peers in a regular guild. He lacked formal education and the knowledge of Latin; he taught in his native language; he wrote down his instructions in his native language. The Indian-Arabic computation by figures was first used by Venetian and Genoese merchants, in the fashion of some esoteric mercantile art. This was suppressed at first by municipal supervisory officials who were not yet conversant with it; Venice, for instance, forbade its use in official account books (1299). The earliest known practical instructions for learning the "Italian" bookkeeping with these numerals date from the 14th century; the oldest ones were prepared by P. Dagomari of Florence (1281?–1365 or 1374). The oldest printed presentations are the anonymous arithmetics of Treviso (1478) and Florence (1481); the best contemporary summary was the book of P. Borghi of Venice (1484). The subject was given thorough treatment also in Luca Pacioli's *Summa* (1494) and in Johann Widmann's *Rechenung* (1489). Coins with numerals imprinted, and printed works with pagination by numbers date from the 15th century.

Among the common people, who could neither read nor write and were ignorant of numerals, too, computation by means of the abacus survived till well into the 17th century. It was taught, in a very clever form, by the *abacists* (the advocates of the older method of computation) under the title (*algorithmus linealis*) "reckoning on the lines" and it was described in numerous printed expositions published in the 15th and 16th centuries. The *algorithmists*, on the other hand, advocated the system of computation by numerals which they had taken, essentially, from the *algorismus* treatises of the Scholastic philosophers (translations and adaptations from the Arabic). Their written methods were still awkward, as for instance, the lengthy method of division upward from left to right, unsuitable for a quick survey, which had originated in counting board computation with easily changeable figures, and which offered no advantage, whatsoever, over the computational methods of the abacists. More appropriate methods, such as the modern left-to-right downward division (at first requiring all the subtrahends to be written out in full, as in the treatise by P. Calandri [1491]) were accepted but very gradually.

Practical requirements were served also by the simple *calendars* of Gutenberg (from 1448) and the improved ones of Regiomontanus (from 1474) with forecasts concerning planetary constellations, eclipses, astrologically important dates and instructions for the calculation of the Christian holidays. There were also manuals with guides for coopers for the gauging of barrels (for instance in G. Reisch's *Margarita Philosophica* of 1503; independently by H. Briefmaler in 1487, by J. Köbel in 1515, and by H. Schreiber in 1518 and 1523). Practical geometry books reproduced the contents of the Roman *agrimensor* manuals. We are acquainted with handwritten compilations of this kind, carrying on the tradition handed on by Gerbert, which were written as late as the 15th century; the first printed presentation is an anonymous *Geometria Deutsch* (1484?).

A characteristic feature of the late Middle Ages, the Renaissance and the early Baroque consists in an unshakable faith in astrology. The former view, based primarily on allegorical parallels, was being displaced ever more and more. The celestial phenomena (eclipses, planetary constellations) which had been drawn into astrology, were regarded now as periodically recurrent processes, capable of being predicted on the basis of

astronomical observations and trigonometric calculations. The
Western Arabic-Spanish planetary tables in current use since
the 13th century, such as the *Toledian Tables* (about 1029–
40; a good explanation of Al-Zarqali) and the *Alfonsine Tables*
(computed by order of Alfonso X of Castile about 1260–66,
and surviving only in a revision made by Johannes de Lineriis
and others about 1300–22), proved to be inadequate. There
was a need for accurate trigonometric auxiliary tables going
beyond the scope of the Western Arab tables, the only ones
known to the West. A re-computation was planned already by
the Viennese master Johannes von Gmunden (1380?–1442); it
was carried out by one of his successors, the uncommonly active Georg von Peurbach (1423–1461) who found an untiring
and reliable collaborator in his brillant pupil, Johannes Regiomontanus (1436–1476). The lectures given by Peurbach on his
planetary theory from 1453–54 on, attracted a great deal of
attention, and Bessarion persuaded him in 1460 to write an
elucidation of Ptolemy's *Almagest;* this unfinished work was
completed by Regiomontanus (1462). Following the sugges-
tions of Peurbach, Regiomontanus set out to write an original
presentation of both plane and spherical trigonometry which
until then had consisted merely of loosely organized laws, theorems and auxiliary tables for astronomical purposes. The first
four books of his opus, *De Triangulis Omnimodis,* contain a
sine trigonometry with many original contributions and important improvements in method as compared with those
taken from his Western Arab and Jewish forerunners (chiefly
Al-Farkhani, Jabir ibn Aflah and Levi ben Gerson), known to
Regiomontanus from the translations by Plato of Tivoli and
Gerard of Cremona. In the later (about 1464) appended fifth
Book, which never became more than a mere collection of
material, use was made (following Al-Battani) of the tangent
function, too; the *Tabulae Directionum* (1464–67) contains
a table of tangents computed to five places, at intervals of one
degree, with decimal subdivision—quite unusual in those days.

Not until 1505 approximately, did the Nuremberg cleric Johann Werner (1468–1528), well versed in all the material then available in manuscript form or in print, get the opportunity to become acquainted with the posthumous work of Regiomontanus. He presented a systematically substantially improved *Spherical Trigonometry* constructed along very original lines (first edition, 1514, surviving only in editions later than 1522) and from 1514 on he applied the so-called prosthaphairetic formula,

$$\sin a \cdot \sin b = \frac{1}{2}\{\sin (90° - a + b) - \sin (90° - a - b)\},$$

which enabled him to reduce multiplication to subtraction.

Modern trigonometry originated in the hands of scholars. In contradistinction to this, perspective was the achievement of artisans and artists. All that was set forth on this subject in the writings on optics by Euclid, Ptolemy and Alhazen, and in the treatises on perspective by Albertus Magnus, Roger Bacon, John Peckham and Witelo, remained at a standstill in its initial stages. Lines converging in depth, in a plane (floor, ceiling) were used, among others, by Giotto (1266–1337) and by Jan van Eyck (1381?–1441), lines converging in space, by F. Brunelleschi (1377–1446). In his *Pictura* (1435, first printing 1511) L. B. Alberti (1404–1472) used a square grid. Piero de' Franceschi (1410?–1492) gave a more specific explanation of the new procedure in his *Perspectiva* (1482–87). His essay on *regular solids* (1487), in which the stereometric computations of domed arches seem to have restored a lost tradition from *The Method* of Archimedes, reached the public as an appendix to Luca Pacioli's *Proportio Divina* (1497, in print in 1509). The perfectly executed illustrations of this text were the work of Leonardo da Vinci (1452–1519) who was, himself, the author of a (now lost) *Perspective*. Albrecht Dürer (1471–1528) became acquainted with the Italian theory of perspective in 1506; he reproduced this knowledge, in a separate chapter of its own, in his *Underweysung*, written for artists (1525). The point of distance method appeared in print for the first time in the *Margarita Philosophica* (1508) by G. Reisch (1475?–1523) and in the *Perspectiva Artificialis* (1505) by J. Pélerin (1445?–1523?).

These interesting, yet on the whole modest beginnings of independent mathematical thinking were overshadowed in importance by the arriving current of new knowledge yielded by closer contact with antiquity. Of far-reaching significance was the fact that the translating activity of the humanists, which was so generously sponsored by Pope Nicholas V (1447–1455), embraced also the exact sciences from now on. To be sure, the translation of Ptolemy's *Almagest* made (about 1450) by the inadequately educated, vainglorious Cretan, Georgius of Trapezunt (1396–1486), was badly done, and was rightfully rejected by Bessarion. Also the translation of a collection of writings of Archimedes made by the cleric Jacob of Cremona (died about 1452) in 1450 had its weak points, but its shortcomings must be forgiven to a great extent because of the novelty of the subject. In 1462, Regiomontanus made a copy of the Latin text, checked it against the Greek original, and provided important improvements; he intended to publish the translation through his printing shop in Nuremberg, but he died before he could carry out this plan. The public was informed of the existence of the Greek original and Jacob's translation through the *Complementa Mathematica* (1454) of Nicolaus Cusanus (Nicholas of Cusa), available in print since 1488. The specimen translation from Archimedes appearing in an encyclopedic work (1501) of the Venetian physician G. Valla (1430–1499) was unsatisfactory. The potential of Archimedean ideas became effective for the first time with the appearance of an original edition, published by Th. Gechauff (1544). Appended to it was the translation by Regiomontanus.

Regiomontanus was the only expert in the Greek language possessing the technical knowledge necessary for the publication of classical texts on mathematics. He wanted to bring into print everything within his reach—writings of Euclid, Diophantus (discovered in 1463), Apollonius, Ptolemy, and several minor authors. His untimely death in distant Rome put an end to all these plans. Euclid's *Elements* was the first classical mathematical text to find its way to the public; a practically unaltered version of the translation of Campanus was published by the Venetian branch of the Augsberg printer E. Ratdolt (1447–1528?), in 1482. The criticism of the humanists was not long in forthcoming; it was directed against the medieval Latin of Campanus, and above all against the differences between the Greek original and the translation. In 1505, B. Zamberti brought out a translation made directly from the Greek, which prompted Ratdolt to publish a revised version of his text (1509). The author was L. Pacioli (1445–1514) who had become most favorably known among professional mathematicians for his *Summa*. Disparities in the wording of the text led to complete chaos, through which J. Lefebre d'Etaples tried to steer by setting in parallel for comparison the Campanus and the Zamberti texts (1516). The editing of Campanus was finally superseded by a Greek edition, published in 1533 by S. Grynaeus (1493–1541). This edition was accompanied by the important commentary by Proclus on Book I of the *Elements*, which was to become more widely known through the (1560) Latin translation by Francesco Barozzi (1538?–1590). But these early editions of Euclid were only partly satisfactory. They still contained factual errors, and errors regarding the view, universally current in the Middle Ages, that the author of the sequential structure of the theorems had been the philosopher Euclid of *Megara* (about 400 B.C.); that the proofs were original with Campanus, making allowance for Theon of Alexandria.

Till about 1540, of Ptolemy's works the *Tetrabiblos* and the *Almagest*, of those of Apollonius the *Conics*, and of those of Pappus, a few excerpts were known; Diophantus was hardly known at all at first. Significant from the mathematical viewpoint were, furthermore, the editions of Plato, Aristotle and their commentators, as well as the Latin editions of the medieval scholars, among whom Boethius, Averroës and Avicenna deserve specific mention. In the domain of mathematics, the late Renaissance was in possession of the most important results of antiquity; the available literary bases were not altogether free from objections as yet, but they did convey a mighty and novel wealth of knowledge and invited the student to delve into the subject more deeply.

Moreover, the 15th century presented the West with two personalities of supreme mathematical genius: Nicholas of Cusa (generally called Nicolaus Cusanus) and Regiomontanus. Cusanus (1401–1464) was no mathematician proper, but a Church statesman and philosopher who tried to unite the Scholastic view with the Neo-Platonist attitude of the Fathers of the Church. He began with the symbolic interpretation of mathematical relationships, as encountered so frequently in his *Docta Ignorantia* (1440), and ended up grappling with mathematical subjects as a true scientist, who, in an objective proof, will admit only facts and logical conclusions based on fact. An enthusiastic humanist, his mind's eye on a meaningful transformation of the no longer tenable ways of life of the last stage of the Middle Ages within the boundaries of the ecclesiastically defined view, he took for his starting point the encompassing power of a *coincidentia oppositorum,* the "concordance of contraries," in which the opposites united and which saw the various well-founded views of one and the same thing as various facets of one and the same truth. He likened knowledge, accessible to Man gradually only, and pure Truth, sealed to the human mind, to the relationships between the inscribed regular polygons and the circle. This led Cusanus to the chief problem in mathematics to which all his scientific experiments were dedicated: the quadrature of the circle. Like Aristotle and Averroës, he was convinced of the impossibility of finding an exact solution of the problem, but he wanted to reach at least an optimal approximation. In the development of the contributions of Thomas Bradwardine on the subject of Zenodorus' treatise on isoperimetrics, Cusanus tried to achieve the quadrature of the circle, $f = r^2\pi$, by the approximate rounding out of isoperimetric regular polygons, $f_n$. He believed at first that it was permissible to assume $f - f_n$ to be proportional to the difference in radii, $r_n - \varrho_n$, of the circumscribed and inscribed circles belonging to $f_n$, but he had to yield to qualified criticism, chiefly to that of P. dal Pozzo Toscanelli (1397–1482), and thereafter he adopted heuristic methods of adjustment which finally led him to the approximation

$$r \approx \frac{1}{3}\,(2r_n + \varrho_n).$$

Cusanus was already familiar with the Archimedean quadrature of the circle; he saw the Archimedes translation by Jacob of Cremona in 1450. He regarded the indirect Archimedean method as the mathematical analogue of his own *coincidentia oppositorum*. His teachings of the *explicatio* and *complicatio* contained the generation of a line by the "flowing" of a point, etc. In this respect he may have been echoing ideas of Sextus Empiricus (*Adversus Mathematicus*) whose writings the Byzantine scholars of that age studied with interest and used in support of the Christian doctrines (as did also Giovanni Pico della Mirandola). In his computations of areas and volumes, Cusanus made use of conclusions involving infinitesimals, becoming the first serious scholar to apply them since the days of the atomists. He still knew little about presenting proof; whatever proofs he does give are based mostly on naive considerations of mean values, and the application of direct proportionality. His writings abounded in unproven assertions which became important problems for mathematicians. His influence was the strongest on those clerics of the subsequent era who were averse to the rigid classificatory logic of Late Scholasticism and who sided with the efforts of the humanists, without giving up allegiance to the view professed by the Church.

Regiomontanus (whose real name was Johannes Müller and was from Umfrieden near Königsberg in Franconia, Germany [1436–1476]) was, unlike Cusanus, strictly a mathematician without philosophical interests, well versed in Greek, but no sympathizer with the cultural endeavors of the humanists. At the age of eleven he entered the University of Leipzig, at twelve he computed, unaided, the planetary movements for 1448, at fourten he left Leipzig and entered the University of Vienna and became the prize pupil of Peurbach. In the years 1461–67 he was in the service of Bessarion in Italy where he made the acquaintance, among others, of the astronomer Giovanni Bianchini (died in 1466) and of Toscanelli, an eminent authority on all fields of natural science and a boyhood friend of Cusanus. During this same period of his life he studied trigonometry, discovered Diophantus and Apollonius, and carried on a correspondence with Bianchini on difficult problems in number theory and problems relating to extreme values, which were taken up as purely geometrical (1463–1464). Both classes of problems were far above the abilities of Bianchini, who was unable to furnish any answer. In 1464, Regiomontanus checked the circle approximations by Cusanus, and after a lengthy and troublesome calculation he found them to be inadequate; but he failed to penetrate to Cusanus' vital thought, which was to be understood functionally only, and therefore his bitter criticism was not quite justified. He was court astronomer of King Matthias Corvinus of Hungary from 1467 till 1471, then moved to Nuremberg to conduct new observations of the skies there and to use his findings to improve the planetary tables which he recognized to be inadequate. He founded a printing shop of his own, which published Peurbach's *Planetary Theory* (1472) and the *Latin* and *German Calendars* (1474), and he formed a plan to publish in Latin translation all the classical Greek mathematical writings known to him. His preliminary activities were interrupted by his trip to Rome (1475) where he was supposed to express an expert opinion in connection with the preparation of the planned reform of the calendar, but he died at the age of forty, at the peak of his creative career. Only parts of his scientifically and historically significant legacy have been preserved, and just a minor portion of those have been published. Whatever of his work is known to this date demonstrates that Regiomontanus is to be regarded as the most gifted mathematician of his century, linguistically able fully to understand the wording of his Greek sources, and scientifically equipped to digest and develop the knowledge gained. It is therefore highly lamentable that the scientific legacy of Regiomontanus was withheld and then partly squandered because of the lack of understanding on the part of his patron, the patrician Bernhard Walther of Nuremberg (1430–1504), so that the life-work of the most important Renaissance mathematician will be accessible merely fragmentarily later on, too.

### 3. THE TRANSITION TO THE BAROQUE (1450–1580)

The dying Renaissance regarded ancient lore still with admiration, but already also with critical reserve. The task at hand was, firstly, a profound understanding of the facts, and secondly, a meaningful and independent elaboration of the regained knowledge. This development took place in an era of direct clashes, often most violent ones, of conflicting worldviews and ideologies (the Reformation, social unrest, the Counter-Reformation). The revelation to the general public of great inventions and discoveries widened the horizon beyond all previous imagination; on the other hand, the well-adjusted bonds within the theretofore strictly regulated social structure kept becoming ever looser. The art of printing enabled the new ideas to be disseminated quickly among the great masses of the people, and thus it became an efficacious tool in the hands of the men who originated the movements of the century that were to change things and conditions. The intense intellectual life of the period comes before us in many diversified forms.

Practical calculation was, for the most part, covered by the instruction given by the reckon masters in the schools maintained for the burgher class; it was treated, of course, thoroughly in the "Latin schools" as well. The system of teaching was based primarily on verbal instruction, memorizing, and mechanical drill. The printed arithmetics were chiefly collections of rules, examples and problems. In addition to a great many German publications, there were also noteworthy Dutch, French, Italian and English presentations of the subject; the Scholastic tradition lingered longest in Spain. The introductions, written in Latin, were still more or less tinted with scholasticism; the best ones among them combined the older knowledge adroitly with the technical advances of the master arithmeticians holding modern views.

These advances related first of all to the treatment of simple equations. Specific individual symbols began to be adopted for the unknown quantity (called *res* in Latin, *cosa* in Italian, *Coss* in German) and its first powers and roots; gradually also, abbreviated operative symbols became generally used, as, for example, $\tilde{p}$ for plus, $\tilde{m}$ for minus; finally, there were true algebraic signs, such as $+$ and $-$. The initiative for this development came from Northern Italy where by 1450 the relevant writings of the Moslems were getting eager attention; Italian, German and French algebraists established the new method using fixed, although still verbose rules which were then shortened more and more. Special mention is due to the *Coss* (1525) by Christoff Rudolff and to its new edition (1553–54) prepared by Michael Stifel, and to Cardan's *Practica Arithmetica* (1539).

The best technical work related to this subject was the *Arithmetica Integra* (1544) by Michael Stifel, a highly gifted algebraist, but inclined to odd numerological superstitions. In that book he stated explicitly, for instance, that negative numbers were smaller than zero. Stifel grasped fully the nature of negative numbers. By admitting also negative coefficients in equations, he was able to reduce the eight principal formulae of the quadratic equation used heretofore, with their twenty-four special rules for solving, to just one single formula. Nevertheless, he still shied away from the recognition of negative roots of equations. With the support of friends versed in foreign languages, who helped him in the analysis of original Greek texts, he fully absorbed the theory of irrationals in Book X of Euclid's *Elements* $\left(\sqrt{a + \sqrt{b}}\right)$ and expanded it by stepping up to expressions of the form $\sqrt[m]{a + \sqrt[n]{b}}$. He treated the numerical extraction of roots up to the seventh root, and the extraction of roots of algebraic expressions in exact examples. The arithmetic triangle was known to him from the title page of the *Arithmetic* (1527) by Peter Apian (1495–1552). He gave the binomial coefficients their name, and he was acquainted with the rule for their construction by addition. He multiplied and divided with the terms $a^k$ of a geometric progression by the adoption of exponents, he extended the progression to include negative exponents, and he presented one to one correspondences such as exist between exponents and powers, viz.:

$$-3 \ -2 \ -1 \ 0 \ 1 \ 2 \ 3 \ \cdots$$
$$\frac{1}{8} \ \frac{1}{4} \ \frac{1}{2} \ 1 \ 2 \ 4 \ 8 \ \cdots$$

This led him to the calculation of powers and roots of magnitudes in geometric progression by the multiplication or division of the corresponding exponents, as well as to dealing with exponential equations. Later scholars, as Robert Recorde (1510?–1558) in the *Whetstone of Witte* (1557) and Simon Jacob in his *Rechenbuch* (1565), were strongly influenced by Stifel; the best comprehensive presentation of the *Coss* appeared in the *Algebra* (1608) of Christophorus Clavius (1537–1612).

In the meantime, about 1500, the Bolognese master Scipio del Ferro (1465?–1526) discovered the algorithmic solution of the cubic equation, presumably through experimenting with cubic irrationalities; but he did not publish his discovery, merely confided it under the seal of secrecy to a few friends and pupils, including the reckon master Antonio Maria Fior (about 1510) and his son-in-law and successor in office, Annibale della Nave (1500?–1558). In 1535, Fior engaged in a public contest with the capable reckon master Niccolò Tartaglia (1500?–1557), in the course of which he proposed thirty cubic equations to his opponent for solution. Tartaglia succeeded in finding the solutions just before the expiration of the time limit and he sent his method, in the form of obscure verses, to Cardan who had pressed him for it more and more insistently and had given assurances under oath, that the secrets would be kept (1539). In 1542, during a visit to Della Nave, Cardan and his pupil, Ludovico Ferrari (1522–1565) ascertained that Tartaglia's method was identical with Scipio del Ferro's. He assumed that Tartaglia had gained possession of that method by dishonourable means, and therefore he considered himself justified in publishing his own method (only slightly different from that of Del Ferro) in his *Ars Magna* (1545), otherwise also important. That book, the printed sheets of which Stifel was able to use, even before the publication of his own *Arithmetica Integra* (1544), also contained among other things, Ferrari's arithmetical treatment of the biquadratic equation $x^4 + px^2 + qx + r = 0$: One had to determine an auxiliary quantity, $t$, so as to make $(2t - p)x^2 - qx + (t^2 - r) \equiv (x^2 + t)^2$ a perfect square. Ferrari's method is related to ways of thinking such as we became thoroughly conversant with in Diophantus. For the first time (since Regiomontanus) we see this kind of thing fraught with meaning, taken up again and carried ahead independently.

Tartaglia immediately protested vehemently against the breach of trust which had been committed, and against Cardan's imputations. He presented his viewpoint at length in the *Quaesiti* (1546), a carefully elaborated arithmetic, but he did not discuss the cubic equation in greater detail, either in that book or in the *General Trattato* (1556–60), a skillful presentation of a complete cross section of the knowledge of the time in the arithmetical and geometrical domains. The controversy over priority in the solution of the cubic equation resulted in a public altercation with Ferrari, which degenerated into a nasty squabble. Of the problems proposed and solved under these circumstances, only a few pretty constructions with fixed compass opening are deserving of attention. In addition, the *General Trattato* contains also a completely individualized solution of an extreme value problem: here, the maximum value of $x(a^2 - x^2)$ was determined in a prescription-like form.

Cardan, who did not participate in the controversy personally, accomplished considerable results in the domain of the theory of equations. In his *Arithmetica Practica* (1539) he had already treated the rationalization of denominators with cubic irrationals; in the *Ars Magna* (1545) he reduced cubic equations to the normal form and investigated the ratios of real quantities in the solution of cubic and biquadratic equations. Cardan was familiar with the relation between the sum of the roots and the coefficients of the cubic equation, and in this case he came close to Descartes' rule of signs. In his posthumous *Sermo de Plus et Minus* (printed in 1663) he treated of the appearance of the imaginary in the cubic equation. This investigation was along the same line as the studies of the Bolognese engineer Raphael Bombelli whose *Algebra* (1572) was devoted chiefly to complex expressions of the form $\sqrt[3]{a + \sqrt{-b}}$. It contains also rules of multiplication for imaginary quantities, algebraic transformations in the irreducible case, and most notably the observation that when Cardan's formula is applied, in the case where the cube roots can be extracted, the imaginary parts must cancel out.

Del Ferro seems to have started out from the question as to when $\sqrt{a + \sqrt{b}}$ can be written as $u + \sqrt{v}$. A reference to this problem appears at the end of Rudolff's *Coss* (1525). Stifel investigated the matter in greater detail in his *Elucidation on the Coss* (1553–54), and similar material appears (although in all likelihood without any knowledge of the prior study) in Bombelli's *Algebra* (1572). The cubic equations solvable by special tricks, such as were introduced by Pedro Nuñez (1502–1578) in the *Algebra* (1564) and by Nikolaus Pieterszoon in the *Arithmetica* (1567), were still unconnected to the general theory of the Northern Italians. This was taken up first by Simon Stevin (1548–1620) in the *Arithmétique* (1585) and by François Vièta (1540–1603) in the *Aequationum Recognitio et Emendatio* (1591) and in the *Supplementum Geometriae* (1593). Both authors made important formal improvements. Vièta mastered, among other things, the irreducible case of the cubic equation, by using a trigonometric way of getting around the imaginary. Interesting variations of this method were presented in the 1615 edition of the *Recognitio et Emendatio,* prepared by Alexander Anderson (1582–1620?), likewise by Albert Girard (1595–1632) in the *Invention Nouvelle* (1629) and by Frans van Schooten (1615–1660) in the Appendix to the *Organica Descriptio* (1646). The *Lustgarten* (1604) by Johann Faulhaber (1580–1635) contained numerous cubic problems, indicating the solutions, but not the methods leading to them. The latter were appended by Peter Rothe (died in 1617) in the *Arithmetica* (1608), with a detailed exposition of Cardan's theories.

In the domain of geometry, the contributions of the practical experts and reckon masters were still insignificant. The *Introductio* of Charles Bouvelles (1470?–1553) was strongly influenced by Thomas Bradwardine (in print since 1495) and Nicolaus Cusanus (in print since 1488); the *Practical Geometry* (1511) shows the influence of Pacioli's *Summa* (1495). Francesco dal Sole (born about 1490) made a noteworthy attempt in his *Arithmetic* (1526) to combine arithmetic and geometry. The development of the science of fortifications provided a powerful stimulus to the study of problems of practical geometry. Let us mention Albrecht Dürer's *Underricht* (1527), the activity in Northern Italy (Verona, 1527) of Michele Sanmicheli (1484–1559), later the theoretical writings of Menno van Coehoorn (1641–1704) and Sébastien le Prêtre de Vauban (1633–1707). The lively interest of the 15th and 16th centuries in architecture also stood the theoretical phases of the subject in good stead: the *Architectura* of Vitruvius Pollio (1st Cent. A.D.) was reprinted several times after 1486, the *Architettura* (1537) of Sebastiano Serlio (died in 1552) was universally studied, and the book on the same subject (1570) by Andrea Palladio (1508–1580) exerted a decisive influence that radiated as far as England. Giorgio Vasari (1511–1574) published his *Vite* (1550) in which he described the lives and careers of the most important architects of his time. Jacopo Barozzi Vignoli (1507–1573), well known to us as a creative architect, presented a new introduction in his *Perspective* (about 1530, at first in manuscript). His *Laws of Architectonic Composition* were considered exemplary by his contemporaries. Barozzi's work was published in 1582 by Egnatio Danti (1537–1586), the mathematician, to whom we owe the Italian translations of several ancient works on *optics* (1573). His annotations contain the first theoretical substantiation of the point of distance rule.

The decisive progress in geometry was inaugurated by the learned students of classical writers on that subject. In their efforts to obtain a text that would be above objection both as to its language and to its content, they kept steadily improving their understanding of their sources. They recorded their findings at first in important editions and translations of those originals. Of particular interest is the Euclid translation (1545) by Pierre de la Ramée, also known as Peter Ramus (1515–1572), who later (1569) in connection with his criticism of Aristotle's system of science (1543), voiced noteworthy objections to the Euclidean method, objections which thereafter would never cease to be heard. The edition brought out by

Federigo Commandino (1572), edited on the basis of Greek
manuscripts and valuable from the point of view of textual
criticism, was soon supplanted by the manual of Christophorus
Clavius (1537–1612) which presented a new Latin version,
stressing the mathematical considerations above all, and giving
detailed explanations and supplementary material. This publication of the celebrated professor of the Roman *Collegium
Germanicum*, author of other valuable writings, was reprinted
again and again during the century and a half following 1574;
it became the authoritative text in the numerous colleges of
the Jesuits, and was used and liked also in other Latin schools
and institutes of classical education; it laid the first secure
foundation of a solid system of education that was capable of
providing the transition to higher subjects.

Those who would go beyond the *Elements,* could find the
best foundation in the Latin editions of the classical writings
prepared by Federigo Commandino (1509–1575). His Archimedes edition (1558) was a promising beginning; the fact that
Commandino fully understood not only the language of the
text, but also its mathematical content, is indicated by the important study on the computation of *centers of gravity* (1565).
To his edition of Ptolemy's *Planisphaerium* (1558) there was
added a short introduction to perspective, which did not, however, go as far as the point of distance method. Of especially
stimulating effect was the edition of the first four Books of the
*Conic Sections* of Apollonius (1566) which provided the first
real insight into the classical theorems of the conics. The ac-
companying supplements from the writings of Pappus, published in full after Commandino's death only (in 1588),
spurred the best minds of the age on to the famous reconstruc-
tions which were destined to lead eventually to the coordinate
geometry and to the theory of extreme values. The Euclid
translation (1572) swept aside at last the old misconception
that Euclid the geometer had been one and the same person
as Euclid the philosopher of Megara (about 400 B.C.) and had

merely formulated the theorems, whereas the proofs had been the contribution of Theon or others.

The mathematical talents of Commandino were excelled by those of Francesco Maurolico (1494–1575), a bilingual Sicilian, some of whose excellent Latin translations with their valuable elucidations were in part to reach the public a century later: his Apollonius translation became publicly known in 1654, his Archimedes translation as late as 1685. His *Arithmetic* (1575) contains an observation of importance in method: In an example, reasoning completely by induction is applied substantially more clearly than in Euclid's books on number theory where it is contained, but hidden, so to speak. Also the last great mathematician of antiquity, Diophantus, was made accessible about this time. In Bombelli's *Algebra* (written about 1560, printed in 1572) we have an excellent Italian revision, based on the Latin translation by the humanist Giuseppe Auria (about 1500). The Latin translation by Wilhelm Holtzmann (1575) was a philologically significant achievement.

The first *independent geometrical treatises* of the Late Renaissance were produced under the influence of the works of classical antiquity. Let us mention first the purely stereometrically oriented study of Johannes Werner (1468–1528) on certain laws of conic sections, then the treatise of Giambattista Benedetti (1530–1590) on constructions with compasses of fixed opening (1553), and finally a few unsuccessful attempts of squaring the circle, as that of Oronce Fine (1494–1555) in 1544, disproved in 1559 by Jean Butéon (1492–1572), and that of Joseph Justus Scaliger (1540–1609) in 1594, which was rejected by François Vièta (1594). The mathematicians of that age placed these quadratures of the circle in the same class as the observation by Jacques Peletier (1517–1582) appearing in his study on Euclid (1557), to the effect that the angle of contingence is actually not a finite magnitude, but equal to zero. This view was rejected by Clavius in particular (in his Euclid edition) ; only Vièta recognized (1593) that Peletier had been right. Almost all the important humanists of the Late Renaissance were very devoted to mathematical study and convinced of its universal significance; I refer specifically to statements to this effect appearing in the writings of Thomas More (1478–1535) and of Juan Luis Vives (1492–1540), but above all to the introductions with which Philipp Melanchthon (1497–1565) prefaced the numerous mathematical editions and writings of German Protestants. His interest was claimed by astrology and astronomy to the same great extent (1553).

This was the very domain in which Nicolaus Copernicus (1473–1543), a man of classical education who had travelled a great deal and far and wide, crowned his investigations of a great many years by formulating the modern view that the Ptolemaic structure of the universe was incorrect and that the Sun, not the Earth, must be placed in the center of the planetary system. A brief *preliminary account* was in circulation in manuscript form since 1514, and from 1533 on a few trusted people knew of the existence of a detailed presentation. In 1539, Copernicus, sought out by Georg Joachim Rhaeticus (1514–1576), mathematics professor in Wittenberg, who absorbed the new lore, was instrumental in having the *preliminary account* published in print (1540), and made the arrangements for the printing of the main work by the Nuremberg printer Johann Petrejus (died in 1551). He supervised the printing until 1542, when that function was taken over by the Protestant theologian Andreas Osiander (1498–1552) who, in view of the unsympathetic attitude of Martin Luther (1483–1546) and Philipp Melanchthon, arbitrarily inserted a preface into *De Revolutionibus* (1543) in which he represented the new teaching—contrary to the conviction of the author himself—as a mere hypothesis. The heliocentric view was rejected at first by the majority of the learned scholars; among its first supporters mention is due to Erasmus Reinhold (1511–1553), composer of the *Tabulae Prutenicae* (1551), a set of tables of planetary movements based on the Copernican system, which soon supplanted the obsolete *Alphonsine Tables*. However, the discrepancies from the observations, caused Tycho Brahe (1546–1601) once again to discard the heliocentric view and to attempt to formulate an adjusted view. The Copernican theory became known in England through the translation of *De Revolutionibus* (1576) by Leonard Digges (died in 1571?), and it gradually supplanted the Ptolemaic system.

*Trigonometry*, too, benefited by the increasing interest in astronomy. In 1533, Johann Schöner (1477–1547) published Regiomontanus' treatise on trigonometry; 1542 was the year of the publication of the Copernican presentation of the same subject, which was further elaborated by Rhaeticus (tables of the trigonometric functions, to seven places, at intervals of 10″) in 1551. The *Canon Foecundus* by Erasmus Reinhold (table of tangents to seven places) was published in 1553, while the great tables by Rhaeticus (1569), computed to 10 places, which did not appear in print till 1596, when Valentine Otho (1550?–1605?) published them, were finally (1613) completed by the tables of sines by Bartholomaeus Pitiscus (1561–1613). Computation with the aid of these many place tables was made substantially easier by the application of the prosthaphairetic methods which were improved almost simultaneously—perhaps under the indirect influence of Johannes Werner—by several scholars.

About the end of the Renaissance, the student of the mathematical disciplines had access to a rich store of new material that invited him to carry on, develop and broaden the achievements extant in every domain. But no great personality had come forth as yet who had the power to survey the entirety, and to combine and unify it. Such a person was the brilliant François Vièta, who must be regarded as the harbinger of a new epoch. The mathematics of the Baroque period begins with him.

## CHAPTER FIVE

# The Early Baroque Period
## (approximately 1550–1650)

### 1. François Vièta (1540–1603)

Vièta began his scientific career as tutor of Cathérine Parthe-
nay (1554–1631) for whom he wrote an elaborate *Astronomy*,
the introduction of which was printed in 1637. He was acquainted with the *preliminary account* (printed in 1540) and
*main work* (printed in 1543 and 1566) of Copernicus; nevertheless, he rejected the new planetary theory because of its
scant agreement with observed facts, and he sought a rational
improvement of the Ptolemaic system. From 1571 on he practiced law, and in 1573 he rose to the post of Crown Attorney.
From 1584 to 1589, temporarily eclipsed by political opponents, he devoted himself entirely to his favorite studies in
mathematics.

The *Canon* (printed in 1571–79) was strongly influenced by
Regiomontanus' treatise on *trigonometry* (printed in 1533)
and by the *Tables* by Rhaeticus (1551), but it was substantially
more accurate than either. Vièta presents, in a skillfully arranged table with twofold entries, the six trigonometric functions to five decimals for every minute, and to ten decimals for
every degree, and he appends the sexagesimal equivalents up
to $k:60^3$; the advantages of the decimal system of notation are
impressively emphasized. By continued bisection of angles, he

obtained from the functions of the angle of 30° those of the angle of $30°:2^{16} = 225':8192$ to 10 decimals, from that he computed $\pi$ to nine decimals, and then, in consequence of $m \sin \dfrac{180°}{n} < n \sin \dfrac{180°}{m}$ ($3<m<n$), the sine of $1'$ to 13 decimals. The methods for the development of $\cos nt$ and $\dfrac{\sin tn}{\sin t}$ as functions of $\cos t$ were not worked out more exactly until later (1593, 1615); spherical trigonometry was developed from the law of sines and a modified form of the cosine law. The rules, today bearing the name of Napier, for the right spherical triangle made their appearance in this connection. For the oblique triangle, Vièta still required the distinction of twenty-eight individual cases; later (in the *Varia Responsa*, VIII, 19) he could get along with six. The excellent approximation $\sqrt[3]{2} \approx \sqrt[4]{2} + \dfrac{1}{20}\sqrt{2}$ and similar considerations warrant the assumption that the approximation method (still clumsy in computation, but important fundamentally) for the *numerical solution of equations* (systematic determination of the next decimal of a given approximate value; printing of 1600) also originated in the early days.

During his banishment, Vièta studied the Commandino editions (printed since 1558) as well as the French editions of the classical works on mathematics, and also the writings of Cardan (1545, 1570), Tartaglia (1546, 1556–60), Bombelli (1572), Holtzmann's Diophantus (1575) and Stevin (1585). The constant repetition in the algorithmic formalism of Diophantus and his commentators led him to introduce letters for numerically undetermined algorithmic quantities, instead of numbers. Thus the *logistica speciosa* was born, in which known quantities were designated by consonants, unknown quantities by vowels.

In the *Isagoge* printed in 1591, powers of the known and unknown quantities were still expressed in words; except for the use of $+$ and $-$, only division was indicated by a symbol (fraction line), multiplication and equality were indicated verbally. The strict observance of the principle of homogeneity seems to indicate the geometric origin of the method, the profusion of Graecisms bears witness to the antiquity of its roots. In his *Notae Priores* (printed in 1631), Vièta constructs and transforms algebraic expressions and also formulates propositions on right triangles which parallel the addition and subtraction formulas of the sine function. In the *Recognitio Aequationum* (printed in 1615), every equation is interpreted—under Italian influence—as a condition existing among the terms of a geometric series. In the *Emendatio Aequationum* (printed in 1615), fractions and roots are removed and easily recognizable factors are separated. Here $x^3 + 3ax = 2b$ is transformed, by means of $y^2 + xy = a$, into $y^6 + 2by^3 = a^3$, and the biquadratic equation is solved by a method connected to Ferrari. In examples involving equations up to the fifth degree it becomes clear that Vièta was able to build up the coefficients of the equations out of the roots of the equations.

In other respects, Vièta had barely risen, as yet, above his Italian teachers. As he restricted the use of letters to positive quantities, he was compelled to make many superfluous distinctions among the individual cases. Leading students of the German *Coss* who had forged ahead in formal theory, such as Clavius, for example, were repelled by the surplusage of unusual technical terms and failed to recognize the rich possibilities of the promising *Logistica Speciosa*. The French, the English and the Dutch, judged otherwise. The new method became more flexible in the hands of Thomas Harriot (1610, in print in 1631), Albert Girard (1629), William Oughtred (1631) and Pierre Hérigone (1634). Modern development began, of course, only with René Descartes (1637).

Vièta wanted to regain the mathematical method which the ancients had wrapped in geometrical garb, but which, in his view, was nevertheless clearly recognizable in Diophantus. This idea prompted him to explore the *Arithmetica* thoroughly, employing algebra as an expedient (*Zetetica*). One noteworthy point was the treatment of undetermined expressions, which were transformed into squares; another one was the reduction of the sum or difference of two cubes into another possible form, $a^3 - b^3 = (a - b^2/t)^3 - (b - a^2/t)^3$ with $3abt = a^3 + b^3$. If the lettered algebra were algorithmically oriented and restricted to numerical examples, it would explain the idea and content of the Diophantine collection, the geometrically oriented, Book VII of the *Collectiones* by Pappus. The general geometric analysis corresponds to the undetermined solution of an arithmetical problem. Everything that was comprehensible algebraically was therefore accessible by means of the *logistica speciosa*.

Accordingly, the *Canonica Recensio* (1591) contains the geometrical interpretation of the fundamental operations and the solution of the equations $x^4 \pm a^2x^2 = \pm b^4$ with compasses and ruler. On the other hand, only quadratic irrationals and their conversions are open to exploration by the method of compass and ruler. But if one admits, as an additional means of construction, the "interpolation" of a given line segment, $A_1A_2 = c$, through a given point $P$, so that $A_1$ and $A_2$ fall onto the straight lines $q_1$ and $q_2$, respectively, then one can (*Supplementum Geometriae*) interpolate two geometrical means (equation: $a:x = x:y = y:b$) and trisect angles (equation: $x^3 - 3x = y$ with $x = 2 \sin t, y = 2 \sin 3t$) and construct the regular heptagon. This purpose requires the solution of the equation $x^3 + x^2 - 2x = 1$ (normal form: $u^3 - 7u = 7$). In general, every cubic or biquadratic equation can be solved by the interpolation of two geometric means and by angle trisection.

Book VIII (the only one published in print) of *Varia Responsa* contains many fine individual topics in a loose sequence. The summation of the geometrical progression $a_1, a_2, \ldots, a_n$ by means of

$$a_1:a_2 = (a_1 + a_2 + \ldots + a_{n-1}):(a_2 + a_3 + \ldots + a_n) =$$
$$(s_n - a_n):(s_n - a_1)$$

is based on Euclid IX, 35. If the absolute value of the quotient is less than 1, then $\lim\limits_{n \to \infty} a_n = 0$ and $s_\infty = a_1^2 : (a_1 - a_2)$. Vièta (like Peletier, who nevertheless remains unmentioned) regards the angle of contingence as zero; consequently, he considers as inadequate the proof of the impossibility of the quadrature of the circle, which was based on the attribute of magnitude in an angle of contingence. He expresses the ratio of the areas, $f_n:f_{2n}$ of the regular polygons of $n$ and $2n$ sides inscribed in a circle of radius $r$ by $2\rho_n:2r$, and he proceeds in an infinite number of steps from the inscribed square, $f_4$, to the circle, $f$, by means of $\dfrac{f_4}{f} = \lim\limits_{n \to \infty} \prod\limits^{\infty} \dfrac{\varrho n}{r}$. Thus he obtains the infinite product of roots

$$\sqrt{\tfrac{1}{2}} \cdot \sqrt{\tfrac{1}{2} + \tfrac{1}{2}\sqrt{\tfrac{1}{2}}} \cdot \sqrt{\tfrac{1}{2} + \tfrac{1}{2}\sqrt{\tfrac{1}{2} + \tfrac{1}{2}\sqrt{\tfrac{1}{2}}}} \cdots$$

as the value of $\dfrac{2}{\pi}$. This is the earliest known purely analytical expression for $\pi$.

An excerpt from the angular sections contains, among other details, the formula

$$\sin t + \sin 2t + \cdots + \sin nt = \frac{1}{2}\left\{1 + \frac{1}{\sin t} + \frac{\cos t}{\sin t}\right\}$$

($nt = 90°$) which is applied directly to $t = 1'$. Thus Vièta stood immediately before the integral formula $\displaystyle\int_0^{\frac{\pi}{2}} \sin t \cdot dt = 1$.

As a supplement he constructs a quadrable circular lune lying wholly on one side of the line connecting their two vertices. He divides one arc into $m$, the other one into $n$ equal parts and states that the construction of the chords corresponding to the arcs will result in the generation of $m + n$ mutually similar circular segments. If the corresponding chords, $2r \cdot \sin mt$ and $2r \cdot \sin nt$, are to each other as $\sqrt{m} : \sqrt{n}$, the problem is solved. Vièta treats of the cases $m = 1$, $n = 2$, 3, 4, and in the latter instance he presents an elegant construction by the interpolation of two geometric means.

Vièta produced the Archimedean spiral $r = a\varphi$ from the circle $r = a \cdot \sin \varphi$ by rectifying the arc $\frac{a}{2} \cdot 2\varphi$; he considers the area $\Sigma$ of the circular segment whose central angle is $\varphi$, as a rectangle on the basis of the expression $\frac{1}{4} \cdot \frac{a}{2} \cdot (a\varphi - a \sin \varphi)$; and he remarks that the tangent at a point of the spiral $P$ ($\Phi$) forms almost equal angles with the chords of the spiral drawn to $P_+(\varphi + \epsilon)$ and $P_-(\varphi - \epsilon)$ ($\epsilon$ being regarded as a constructable part of 90°). If $\varphi = 90°$ and $\epsilon = 90°$, he finds a difference between the two angles of about 45′; when $\varphi = 90°$, $\epsilon = 4\frac{1}{2}°$, the difference is not more than $\frac{1}{10}′$. Thus, the difference between the angles diminishes boundlessly as $\epsilon(> 0)$ diminishes. We see in this reasoning the first attempt to pass directly to a limit, in a case by no means trivial.

The publication of Book VIII of the *Responsa* (in the early part of 1593) took place during a controversy with the learned Joseph Justus Scaliger (1540–1609), a man of great and diversified knowledge, who made a great to-do about allegedly having found constructions for the quadrature of the circle (involving $\pi = \sqrt{10}$); for the duplication of the cube, and for the trisection of the angle with compasses and ruler, and who referred to his critics brusquely as ignoramuses. As he was no match for Vièta's criticism, he left France (in 1594) to accept a professorship in Leyden. His *Cyclometrica Elementa* and *Mesolabium* published there (1594) were disproved at once by the most capable contemporary mathematicians. The most interesting of the treatises in opposition were the two refutations written by Vièta.

In *Munimen Adversus Cyclometrica Nova* (1594) Vièta approximates a circle segment by means of the segment of an inscribed spiral of Archimedes, tangent at one vertex, and arrives at a relationship equivalent to $\dfrac{3 \sin t}{2 + \cos t} < t$; this proves the previously mentioned rule of Nicholas of Cusa (cf. pp. 78–79) to be a mere *approximation*. Vièta was perhaps not aware of this connection, but he knew that a similar approximation was feasible by means of a parabolic segment, too. His rule stating that the circular segment is greater than $4/3$ of the inscribed isosceles triangle leads to $\dfrac{\sin t \,(4 - \cos t)}{3} < t$ and is only a little more inexact than the previous approximation. This method of approximation was destined to be developed in an amazing form by Christian Huygens (1629–1695) later (1654).

In *Pseudomesolabium* (1595) Vièta studies those chords of a circle, which intersect the diameter in such a manner as to produce four segments in a geometric series. In an appendix, he refutes the assertion by Scaliger that in a quadrilateral inscribed in a circle the diameter is the arithmetic mean between the two diagonals. Influenced perhaps by calculations appearing in the correspondence of Regiomontanus with Bianchini (1464), circulated in those days in manuscript form, he makes use of Ptolemy's theorem, $ef = ac + bd$, and of the formula $e{:}f = (ad + bc){:}(ab + cd)$, to develop, from the four sides, $a$, $b$, $c$, $d$, the expressions for the diagonals, $e$ and $f$. His construction is based on the fact that, for instance, for the angles $\alpha$, $\beta$ between the sides $d$, $a$ and $a$, $b$, the following equation holds true: $\sin \alpha : \sin \beta : \sin (\alpha + \beta) = (ab + cd){:}(ad + bc){:}(a + c)\,(a - c)$.

It is improbable that Vièta was acquainted with the integral quadrilateral inscribed in a circle mentioned in Simon Jacob's *Rechenbuch* (1565), where $a = 25$, $b = 33$, $c = 60$, $d = 16$, with the diagonals $e = 52$, $f = 39$, and the diameter of the circumscribed circle $2r = 65$; the fine study of Giambattista Benedetti (1530–1590) in the *Speculationes Diversae* (1585), too, as well as the constructions in the *Problema Geometricum* (1586) of Francesco Barozzi (1538?–1590?) seem to have been totally unknown to him.

A few years later (1598) the quadrilateral inscribed in a circle was treated in detail in an independent study devoted to it by Johannes Richter (1537–1616) who taking his start from Vièta paid due regard also to the earlier contributions of the Germans; we are indebted to him for a method of general validity for the construction of quadrilaterals inscribed in circles, with integral sides and diagonals, where its area and the diameter of its circumscribed circle also, become integral. Richter had heard also that other mathematicians of renown had worked on the problem in question; in addition to Benedetti, this may have referred to Bombelli who—in that part of his *Algebra* (about 1565) which never got beyond the manuscript stage—had calculated the diameter, $x$, of the circumscribed circle directly on the basis of the relation

$$a \sqrt{x^2 - b^2} + b \sqrt{x^2 - a^2} = c \sqrt{x^2 - d^2} + d \sqrt{x^2 - c^2}.$$

Vièta is the basis also of the investigation contained in the posthumous *Fondamenten* (1615) by Ludolph van Ceulen (1540–1610) who related his construction to the calculation of the segments produced by the prolongation of two opposite sides of the quadrilateral. Formal improvements were added by Willebrord Snell (1580–1626), translator and editor of the *Fondamenten*. He obtained, by induction, from the general formula of the triangle, the area formula $\sqrt{(s-a)\ (s-b)\ (s-c)\ (s-d)}$, which was frequently mentioned thereafter, for instance, in the *Quaestiones* (1618) by Benjamin Bramer (1588–1650?) and in the *Trigonométrie* (1626) by Albert Girard (1595–1632). But the first to be able to prove it (1727) was Philipp Naudé (1684–1747).

Vièta acquired fame all over Europe as a result of his contest with Adriaen van Roomen (1561–1615). Van Roomen presented a survey of the most important living mathematicians of the era in his *Ideae Mathematicae* (completed in 1590, printed in 1593), but he did not mention one single French mathematician, not even Vièta. With reference to that publication, the ambassador from the Netherlands to the court of Henry IV in Fontainbleu expressed himself scornfully concerning the scientific achievements of the French, saying that there were certainly none among them who could solve the problem propounded, under promise of a prize to the solver in the *Ideae*, which problem required the solution of an equation of the forty-fifth degree (the division of an angle equation) under certain special conditions. Vièta, summoned to meet the challenge, gave one solution immediately, and the next day he supplied the other twenty-two (positive) solutions of the problem. In his *Responsum* (1595) he summed up his method of solution, corrected an error in the formulation of the prize problem, and indicated its connection with the general equations for the division of angles into three and into five equal parts. He repeated the remarks made in the *Varia Responsa* about angle sections; moreover, he actually set forth the formulae for the chord $y_n = 2 \cdot \cos nt$ in the unit circle from the basic chord $x = 2 \cdot \cos t$. The generating recurrence formulae $y_{n+1} = xy_n - y_{n-1}$, appeared (in a slightly less definite form) only in the posthumous edition of the *Sectiones Angulares* (1615), brought out by Alexander Anderson (1582–1620?) with added proofs. That publication also expressly stated the relationship between the array of coefficients ("*Canon*") and that of the arithmetical triangle.

At the end of the *Responsum*, Vièta—with reference to the lost *Tactiones* of Apollonius—proposed to the challenger the problem of constructing circles tangent to three given circles. Van Roomen knew of a declaration by Regiomontanus that he had despaired of a solution with compasses and ruler. Therefore, he determined the centers of the circles sought as the intersections of two hyperbolas (1596). Vièta pointed out the possibility of an elementary solution, with reference to Pappus

(*Collectiones* VII), as early as 1597; in 1600, he published his method, taken from his earlier papers (Book VI of the *Varia Responsa*).

The last years of Vièta's life were filled with fruitless attacks on the reformation of the calendar (1582) by Gregory XIII (Pope from 1572 to 1585), coupled with violent assaults on Clavius, one of the advisers of the Pope in this matter. Vièta's point of view was untenable; the caustic presentation of his arguments embittered the adversaries and spurred them on to an overhasty rejection of the incorrectly evaluated *logistica speciosa*.

A small collection of Vièta's writings was published, in many an instance with excellent elucidations by Anderson, in 1615. Marin Mersenne (1588–1648) struggled for years to assemble a complete edition of all of Vièta's published work; in 1637–38 he hoped to win the energetic Elzevier of Leyden over as publisher, and he allowed P. de Fermat (1601–1665) to examine some of Vièta's posthumous treatises to check their suitability for printing. In 1641, he found an interested publisher in the person of Schooten, who published the submitted material in a carefully edited version and with occasional simplification of the symbolism used (following the Cartesian system of notation). Only through this edition did the writings of Vièta become widely known. Many a point was obsolete by 1646, many others were not understood properly, notably the contributions to analysis. The most valuable portions of Vièta's posthumous works, namely the first seven books of the *Varia Responsa*, seem to be lost. A few algebraic works, which Fermat regarded rightfully as already unimportant, still exist in manuscript only.

## 2. THE CONTEMPORARIES OF VIÈTA AND THEIR PUPILS (1550–1650)

In the eyes of the contemporaries of Vièta, who—except for a very few—hardly knew of him at all, the representatives of the Dutch school (becoming more and more powerful after 1580 or so) were the leaders in the domain of mathematics. The most important personality among those Netherlanders was the military engineer Simon Stevin (1548–1620) of Bruges, who appeared in Leyden after travelling about for ten years. In his *Tables for Computing Compound Interest and Annuities,* printed in 1582, he tabulated $(1 + q)^{\pm n}$ and $\Sigma (1+q)^{\pm x}$; in his *Geometrical Problems* (1583) he treated of several semi-regular solids, carrying on the ideas of Dürer (1525). Calculation with decimal fractions and its practical significance is the subject of the frequently reprinted *De Thiende* (1585) which was also appended, under the French title *La Disme,* to the French *Arithmétique* (1585). Stevin presented here the best from older arithmetical and cossistic writings (for instance from Gemma Frisius [1540], Stifel [1544], Bombelli [1572]), independently elaborated and formally improved by him. Although he still had no general symbols for the coefficients of equations, he admitted both positive and negative coefficients and roots of the equations; he recognized irrational quantities as numbers, and knew of the difficulties involved by the appearance of the imaginaries. He denoted powers of the unknown by encircled additional numbers, and represented also fractional powers in this fashion—although that was a mere matter of notation, without any operational application as yet. Otherwise he held fast to the pure arithmetical conception of the arithmetical operations. A noteworthy contribution by Stevin is his rule of approximation for the solution of higher equations; it amounts to a procedure of repeated trials with consecutive decimal subdivision. Particularly interesting is the annexed French translation of the first four books of the *Arithmetica* of Diophantus. Stevin followed Holtzmann's text (1575) and made use of Bombelli's studies (1572) appertaining to it. He contributed valuable improvements in meaning, and had a stimulating influence on the later editor, Girard, who—for instance—was led (1634) to the assertion that every prime number of the form $4n + 1$ can be expressed in exactly one way as the sum of an even and an odd square number. The definition of the ellipse based on the property of its focal points, appearing in the *Geometry* (1585), was originated by Monte (1579).

The friends of Stevin included the master Rudolph Snell (1546–1613; from 1581 at the University of Leyden, founded in 1575), from 1585 Ludolph van Ceulen, and from 1593 Scaliger; among his correspondents mention is due to Adriaen van Roomen. Stevin introduced Prince Maurice of Orange (1567–1625) to pure mathematics, and to practical mathematics in particular (bookkeeping, physics, engineering, military science) in 1582–83, and later the Prince placed his former teacher in important military posts of trust. Beside the important technical inventions of Stevin (sailing craft, improvements in windmills and sluices), we must not forget his *mechanical-hydrostatic works* (1586). They carried on the ideas of Archimedes and Heron and of the works (still considerably lacking in independence) of Commandino (1565) and Monte (1577). The starting point was constituted by the law of the lever, the determination of the center of gravity, and the buoyancy theorem. Stevin treated the inclined plane in an original way, and he advanced to the parallelogram of forces, to the concept of the metacenter, to the paradox of hydrostatics, and to the determination of the upward, downward and sideward pressures. In the attempt at a direct application of the Archimedean method, he made use purely arithmetically of

$$s_n = \sum_{k=1}^{n} k/n^2 = \frac{1}{2} + \frac{1}{2n} \text{ and } s_{n-1} = \frac{1}{2} - \frac{1}{2n} \text{ and prepared for the}$$

transition to the limit: $\lim_{n \to \infty} s_n = \frac{1}{2}.$

In the widely read *Mémoires Mathématiques*, printed in a Dutch, a French and a Latin version (1605–08), he presented, beside many monographs of a special character (for instance, on double-entry bookkeeping in the handling of public revenues), a noteworthy treatment of perspective which excels its prototype (Monte, 1600) in many a respect, and also a study on the loxodromic curve. This curve appeared on the globe prepared by Gerard Mercator (1512–1594) for Emperor Charles V (1519–56) in 1541, and it was studied in greater detail by Nuñez (1546). Stevin constructed templates to fit the sphere for drawing this curve; the technical term "loxodrome" was introduced by Willebrord Snell in the Latin translation of the *Mémoires*. The appended tables were taken from the *Navigation* (1599) by Edward Wright (1558–1615). They show the distance $v$ of a point of geographic latitude $\varphi$ from the Equator on the map, using Mercator's projection (a cylindrical projection so applied that the increase in latitude $v$ on the map becomes proportional with the modified increase in latitude $\varphi : \cos \varphi$ on the globe): $v \sim \Sigma \triangle \varphi : \cos \varphi$.

Stevin was a practical man, well versed in all pertinent methods of computation and construction, and yet by no means lacking in theoretical knowledge and understanding. The interest in problems of application in connection with nautical, military and geographical questions manifested itself among his contemporaries in many small improvements in the domain of trigonometry. The function designations *sine, tangent* and *secant,* suggested by Thomas Fink (1561–1656) in 1583, gained general acceptance very rapidly; abbreviations and functional symbols soon began to be used in general transformations and practical calculations, but no final form was reached as yet. The formalism in the treatment of plane and spherical triangles was becoming simplified; in this connection, the prosthaphairetic method, based on the formula $2 \cdot \sin a \cdot \sin b = \cos (a - b) - \cos (a + b)$, served to reduce multiplication to subtraction. This method, devised by Werner (1514), was inherited from him by Rhaeticus (1542), and the latter passed it on, through Richter (1569), to Brahe who—with his pupil, Paul Wittich (1555–1587)—from 1582, made the new method the basis of practical trigonometry. Burgi became acquainted with prosthaphaeresis. N. Rymers (died 1599), probably initiated in it by Brahe and Bürgi, published the method unrightfully as his own invention in 1588. Clavius brought out a definitive presentation in 1593. This well developed method of computation remained current for more than half a century and was supplanted only slowly by the simpler and more efficient logarithmic method.

In the domain of the *theory of quadrilaterals,* only simpler problems were treated, as resection in the three point problem, utilized already by Augustin Hirschvogel, municipal architect of Vienna. A detailed description of the technique was presented by Willebrord Snell (1617); the current practice of naming it for Laurent Pothenot (1692) is unjustified. Snell's *Trigonometry* (1627) treated also of resection in the two point problem long before Peter Andreas Hansen (1795–1874), whose name the problem has borne since 1841. The best tables of the era were composed by Bartholomaeus Pitiscus (1561–1613) whose *Thesaurus* (1613) contains sine values to fifteen places at intervals of 10″, together with the differences of first, second and third orders—only about 0.3% of the indicated values are incorrect.

In connection with the formal development of trigonometry and with the computation of tables to many places, there was a demand for an ever greater accuracy in the calculation of $\pi$. The clumsy attempts of Simon Duchesne (1583, 1586), Joseph Justus Scaliger (1594) and Christian Severin Longberg (1562–1647) were rejected by numerous mathematicians at once. The excellent approximate value $\pi \approx \frac{355}{113}$ (6 correct decimals) was produced, about 1573, by Valentin Otho (1550?–1605?) and independently by Adriaen Anthonisz (1543?–1607) from the approximations $\frac{22}{7}$ (Archimedes) and $\frac{377}{120}$ (Ptolemy) through subtraction of the numerators and denominators, and it was correctly judged as to its merit. The best approximations for $\pi$

to many places were worked out by the Netherlanders. They applied Archimedes' method at first, and reached their goal only after tedious calculation of continued square roots. Van Roomen found fifteen decimals from the polygon of $15 \cdot 2^{24}$ sides in 1593; Ludolph van Ceulen (1540–1610) obtained twenty decimals from the polygon of $15 \cdot 2^{37}$ sides in 1596, later thirty-two decimals from the polygon of $4 \cdot 2^{60}$ sides (printing in 1615), and finally thirty-five decimals (report of Willebrord Snell, 1621).

The newer development began with Willebrord Snell (1580–1626) who as a young man of twenty had given public lectures on Ptolemy in Leyden and during his extensive travels made the acquaintance of Roomen in Würzburg and of Brahe and Kepler in Prague. His first publication, a reconstruction of the *Determinate Sections* of Apollonius on the basis of a survey by Pappus (1608), still in Greek, and unpublished, is significant mathematically, a failure historically. The Latin translations of Stevin's *Mémoires* (1608), those of the writings of Ludolph van Ceulen on the quadrature of the circle (1615, 1619), and the publication of the Latin *Arithmetic* of Ramée (1613) enriched the originals by valuable supplements. In *Eratosthenes Batavus* (1617) Snell outlined his mensuration of degrees; it is not known when he discovered the law of refraction (report of Gool, 1632). In *Typhys Batavus* (1624), he investigated the loxodrome, elaborating further on the findings of Stevin (1605–08).

In calculating $\pi$ to thirty decimals from the inscribed polygon of $4 \cdot 2^{44}$ sides, in 1616, Philipp van Lansberge already used an empirically established approximation, equivalent to $t < \dfrac{\pi \sin t}{2 + (n-2)\cos t}$. In the *Cyclometricus* (1621), Snell used, instead, a combination of inscribed and circumscribed regular polygons which yields $\dfrac{3 \sin t}{2 + \cos t} < t < \dfrac{1}{3}(\operatorname{tg} t + 2 \sin t)$. It is questionable whether or not he was aware of the equivalence of the lower limit with the rule of approximation of Nicholas of Cusa (1458), but he certainly was acquainted with the rule of Vièta (1594). Snell was unable to produce the desired elementary proof for his approximations; that was first accomplished by Huygens (1654) who found new and better rules.

The value of $\pi$ to thirty-five places was used by Jan Storms (1559–1650) for the continued fraction expansion of $\pi$ and $\sqrt{\pi}$ (1633). This involved the use of a method of calculation the first traces of which had appeared in Bombelli's work (1572),

viz.: $3\frac{2}{3} > \sqrt{13} > 3 + 4:\left(3 + 3\frac{2}{3}\right).$

In dealing with special examples of square roots (1613), Pietro Cataldi (1548–1626) performed calculations in which step by step continued fractions were generated, and he recognized the approximation from both sides. Independently of this, Daniel Schwenter (1585–1636) indicated a purely arithmetical method based on continued division for approximating fractions expressed in integers, to fractions expressed in smaller numbers (1618–1636). His exposition seems to have been Storms' source material.

The major part of Storms' work belongs to an oracle on dice, consisting of an arrangement of certain prophetic verses on the twenty-one different throws of two dice. Such oracles ("books of chance") were universally current in those days; a very old one was authored by Heinrich Vogtherr (1539). Related subjects appeared also in contemporary mystical numerological writings, as in those of Jodocus Clichtovaeus (1513) and Pietro Bongo (1583–84). More profound were the *Magical Writings* (1510 and 1531–33) of Cornelius Heinrich Agrippa von Nettesheim (1486–1532). A curious word numerology was written by Stifel (Appendix to the *Bearbeitung der Coss*, 1553) and by Johann Faulhaber (*Calendar* for 1618), culminating in forecasts of coming events. The fact that genuine mathematical knowledge may spring from such superstitious ideas is demonstrated by the theory of *magic squares* which reached the West in the 16th century where it was greatly encouraged. Particularly instructive is the "terrace method" (*méthode des terrasses*) for squares of an odd number of cells, set forth in the *Problèmes Plaisans* (1612 and 1624) by Claude Gaspard de Bachet de Méziriac (1581–1638).

Bachet's book was the first important collection in the domain of mathematical recreations, and it goes far beyond the older literature. It contains the first rule for the solution of the equation $ax + by = c$ in integers, and it shows how two expressions of the form $ax + b$ can be jointly made into squares. Equally valuable are his explanations of the Greek-Latin edition of Diophantus (1621), in which he gives careful consideration also to the findings of his predecessors. Bachet was used as a source by Girard (Stevin editions of 1625 and 1634) and likewise by Pierre de Fermat (1601–1665) who became the founder of modern theory through generalization, transformation and deepening of theories and teachings based on Diophantus.

In the domain of algebra, the use of letter symbols slowly supplanted the symbolism of the cossistic system. A particularly important part was played by the division of an angle problem, which made the introduction of general notations of powers necessary. This problem was treated (in connection with the Archimedean polygon method for the mensuration of the circle) above all in the Netherlands. We must mention, also, on the one hand, the computation of the sums of powers of consecutive numbers by Johann Faulhaber (1580–1635), a man inclined to abstruse numerological superstitions, who went up to $\Sigma n^{17}$, and, on the other hand, the calculation of the sums of the lowest powers of the roots of an equation (up to $\Sigma x^4$) by Albert Girard. Mention is due, finally, to such beginnings as aimed in the direction of the fundamental theorem of algebra: Peter Rothe recognized in 1608 that an equation of the $n$th degree has at most, $n$ solutions; independently of this discovery, Thomas Harriot spoke (1610, in print in 1631) of $n$

linear factors of a polynomical of the $n$th degree, and Albert Girard declared in 1629 that, including the complex solutions, every $n$th-degree equation has exactly $n$ roots. Theoretically significant thoughts of this kind, however, appeared only sporadically at first; practical interests were predominant.

Similarly, this holds true for the many presentations devoted to the developing theory of *perspective*. Monte (1600) is deserving of special emphasis: Using the reflection method (turning the figures over), he proved the concurrence of the images of parallel straight lines at the vanishing point, and he determined the position of the beholder's eye from the given perspective of a straight line. In 1631, Vaulezard tried to construct the object from several perspectives. Jean François Nicéron (1613–1646) went into a more detailed study of the appearance of distortion in perspective representations in 1638, and Pierre Alleaume established the scale of vanishing points in the representation of horizontal straight lines (1628, printed in 1643). This provoked a contradiction by Girard Desargues (1591–1661), who advocated construction using the point of sight alone (1636, 1643). Violent controversies ensued in consequence of which Abraham Bosse (1611–1678), an enthusiastic follower of Desargues, had to resign his professorship at the Paris Academy of Arts (1655).

The numerous geometrical treatises of this stage of the development were also concerned primarily with practical applications (for instance, the measuring of tracts of land, and the computation of the capacities of barrels). They brought about a greater familiarity with geometrical problems formerly rather neglected, and prepared the ground for the progress that was to follow. There was a marked trend toward an increasingly more intimate fusion of the algebraic and geometric points of view.

Neat elementary quadratures of figures bounded in part by circular arcs, had already appeared in Leonardo da Vinci's (1452–1519) writings, and later in Vièta's, (1593). Giambattista della Porta (1538–1615) published a collection of examples of this type, in 1601; he observes that a concave-convex triangle bounded by two arcs and a line segment is quadrable by elementary methods if the ratio of the central angles of the arcs is 2:1, and the ratio of their radii, $\sqrt{2}$:1. He seems to have re-discovered, independently, the quadrature of the lune in the form of a general right triangle which Ibn Al-Haitam (965?–1039) had discovered several centuries before him. Paolo Aurineto presented an elegant summary of Porta's chief accomplishment in 1637. Similar information appeared in a book for young readers, authored by Artus de Lionne (1583–1663) in 1610, but not published in print until 1654 (and which, incidentally, makes no mention of Porta).

The achievement of greatest consequence of the late 16th century in the domain of practical mathematics consisted of the invention of *logarithms* by Bürgi (from 1588) and Napier (from 1594), and in the change to the common logarithms (i.e., those using 10 as base) proposed by Briggs (1615).

Jost Bürgi (1552–1632), a Swiss watchmaker and toolmaker who knew no Latin at all, applied the prosthaphairetic method, expediently developed by the introduction of auxiliary angles, to the calculation of spherical triangles, about 1586; about 1592, he expressed the chords of multiples of arcs analytically ($x = 2r \sin t, y = 2r \sin nt$; taking integral values of $n$ up to 20) and he solved the resulting equations by skillful computation (variations of the *regula falsi*). Influenced by the remarks of Stifel about the multiplication and division of terms of a geometric progression by the addition and subtraction of their exponents (1544), with which he was familiar from Jacob's *Rechenbuch* (1565), he clearly saw the fundamental basis for computation by logarithms, and (about 1600) he calculated his *Progresstabulen* (printed in 1620), a table of antilogarithms. Starting out from $y = 10^8 \cdot 1.0001^x$ ($x$ given to 4 places, $y$ to 9 places) and proceeding by step-by-step multiplication and skillful interpolation, he found a correspondence between $x =$

230,270.022 and $y = 10^9$. He seems to have used higher differences in the practical computation, as well as in his lost table of sines (in multiples of $2''$). The base of this system would be $1.0001^{10,000} = 2.71846 \approx e$.

The tables of Bürgi were published under unfavorable external circumstances (November 8, 1620, the conquest of Prague), and they remained almost unknown. Luck was kinder to John Napier (1550–1617) who drew directly on Euclid and Archimedes. As indicated by his table (*Descriptio*, printed in 1614) and by the posthumous *Constructio* (printed in 1619), Napier started out from a mechanical consideration, to which he may have been led by the special case rule proposed by Thomas Bradwardine: A point $P(x)$ moves along a straight line with the constant velocity $\dfrac{dx}{dt} = c$, while simultaneously, a corresponding point $Q(y)$ moves in the opposite direction, with the velocity $\dfrac{dy}{dt} = \dfrac{y}{b}$ depending upon $y$. From this it becomes clear that the Naperian logarithms are, fundamentally, logarithms of base $1/e$. In working out his system, he paired the terms $x_n = n(1 + 0.5r)$ of an arithmetic progression with the terms $y_n = (1 - r)^n/r$ of a geometric progression ($r = 10^{-7}$). The system of logarithms characterized by this arrangement would, after reduction, have the base

$$(1 - r)^{1/r(1 + 0.5r)} \approx \frac{1}{e}\left(1 - \frac{1}{3}r^2\right).$$

Napier's table contains the logarithms, to seven places, of the sines and cosines at intervals of one minute, and the corresponding differences (lg tg). For numerical purposes, Napier used 10 mutually adjustable calculating rods, with the small multiplication tables marked on them (1617). This tool, highly praised by his contemporaries, was improved by Kaspar Schott (1606–1666) who substituted a rotating cylinder for the rods.

Napier's tables were received immediately with the greatest interest. Benjamin Behr (1587–1633) had an abridged reprint brought out in 1618, and in 1624 he published a re-calculation, with the correction of a newly discovered computational error in the original version. Kepler became acquainted first with Behr's table (1618), then with Napier's (1614), and he carried out a finely planned re-computation (1624–1625) which was continued by his son-in-law, Jakob Bartsch (1600–1633), in 1630–31. Peter Crüger (1580–1639) published a table (in 1634), based on Kepler's *Rudolphine Tables* (1627), in which the numerical logarithms were separated from the trigonometric logarithms. An appendix, written probably by Oughtred (1574–1660), to the second English edition of the *Descriptio* (1618), already laid the groundwork for the determination of the boundary value $\lim\limits_{n \to \infty} \left(1 + \frac{1}{n}\right)^n$ and stated a modest number of natural logarithms; the first 1000 natural logarithms were listed by John Speidell (1622).

Henry Briggs (1561–1630) was the occupant of one of the seven chairs of science founded in London by Th. Gresham (1519–1579). Thereafter, as a professor at Oxford he became the first occupant of the Chair of Astronomy founded by H. Savile (1549–1622). He studied the *Descriptio* critically from beginning to end. In 1615, he proposed the simplifying assumption, $\log 1 = 0$, $\log 10 = -1$, and compromised with Napier on the assumption, $\log 1 = 0$, $\log 10 = 1$. His newly calculated tables (1617: 14 place logarithms of numbers from 1 to 1000; 1624: 14 place logarithms of numbers from 10,000 to 20,000 and from 90,000 to 100,000; 1628: 10 place logarithms of numbers up to 100,000, completed by E. de Decker and Adriaen Vlacq), were accepted—in most instances with minor changes and abridgements—by Edmund Gunter (1620), Edmund Wingate (1625), Denis Henrion (1626), Johann Faulhaber (1630), Nathaniel Roe (1633), and many others. Gunter's tables of angles, using the sexagesimal division of a degree employed up to this time drove the centesimal division preferred by Briggs (1633) into the background. The adroit adaptation of the trigonometric formulae to logarithmic calculation, and the use of higher tabular differences, are two noteworthy features of this work, published by Henry Gellibrand (1597–1637). Logarithms did not become fully accepted in practical computation immediately; that required about two generations. In the transition period, trigonometric problems were treated both prosthaphairetically and with the aid of logarithms, as, for instance, by Georg Ludwig Frobenius (1634).

In calculating their tables, both Bürgi and Napier applied rather complicated, purely rational methods. To calculate a logarithm, of base 10, Napier suggested (1619), as did Butéon (1554: the approximation, $>\bar{k}$) the interpolation of $2^k$ geometric means between 1 and 10, from which he computed even closer approximate values for the logarithm through successive extraction of square roots and multiplication. He noted, furthermore, that, from the number of digits $n$ in $a10^p$, it followed that $\lg a \approx (n-1):10^p$. In 1624, by practical computation, Briggs constructed fifty-four numbers of the form $\sqrt[n]{10}$ ($n = 2^k$), calculated to 32 decimals, to be combined by multiplication, for the antilogarithm $a$ (between 1 and 10). Besides this, he used a unique resolution of $a$ into factors of the form $1 + c_k10^{-k}$ ($c_k$ integral, between 1 and 9; $k = 1, 2, 3 \ldots$). For the approximate computation involved in the repeated extraction of the square roots of $1 + x$ ($|x|$ small), he made use of a polynomial array of coefficients which he represented rationally from the last term (anticipation of the binomial expansion for $\sqrt{1+x}$). He carried this development, taking $n = 2^{-k}$, as far as $\frac{10^n - 1}{n} \approx \frac{\sqrt{10^n}-1}{n/2} = M$, from which he derived, for $\sqrt{10^n} < 1 + x < 10^n$, the relation $M = x/\lg(1+x)$, and from that the expression $\lg(1+x)$. For the computation of $\lg 2$, he started out from $2^{10} = 10^3 \cdot 1.024$; for larger prime numbers, from

$$\lg x = \frac{1}{2}\left\{\lg \frac{x^2}{x^2-1} + \lg (x+1) + \lg (x-1)\right\}.$$

He interpolated (1633) under the assumption of constant 20th differences. We learn on this occasion that Thomas Harriot (1560–1621) had determined the area of a spherical triangle as early as 1603.

Later, Nicholas Mercator (1620–1687) developed a rational procedure for the computation of logarithms, which he presented in the first part of his *Logarithmotechnia* (1668); this could have been derived by transformation from Napier's method. Now, Briggs' method of differences was skillfully developed. Finally there appeared the very close approximation formula, $\sqrt[n]{\dfrac{a-x}{a+x}} < \dfrac{an-x}{an+x}$ $(0 < x < a, n > 1)$ by means of which, for values of x sufficiently small, logarithms could also be determined.

Gunter constructed a logarithmically graduated calculating scale about 1620 (description by Wingate, 1624). Oughtred began to use rectilinear and circular sliding scales in 1622 (printing in 1632–33). His pupil, Richard Delamain, improperly claimed the circular arrangement as his own invention in 1630. The slide rule with an inserted slide was used by Wingate (1654) and Seth Partridge (1662). Johann Ciermans (1602–1648) is said to have constructed a calculating machine as early as 1640; the first known adding and subtracting machine was invented by Blaise Pascal (1623–1662). Gottfried Wilhelm Leibniz (1646–1716) was the creator of the first graduated rolling machine, capable of handling all four fundamental arithmetical operations, automatically, in the same way (1671; first operating model, 1674; description, 1694). Leibniz' machine was far superior to the apparatus by Samuel Morland which was constructed out of a combination of Napier's rods and Gunter's scales.

### 3. On the Way to New Discoveries
### (1550–1650)

The humanistic ideal of learning brought a change in the general view of the nature and task of education, which in turn resulted in an especially intensified cultivation of the ancient languages and manifested itself in particular in the epistle *De Formandis Studiis* of Rudolf Agricola (1443–1485), in the treatise by Desiderius Erasmus (1467–1536) entitled *De Ratione Studii,* and in the inaugural lecture given by Philipp Melanchthon (1497–1565) under the title *De Corrigendis Adolescentiae Studiis.* Melanchthon, who had espoused the anti-peripatetic attitude of Luther (1520–22), returned to Aristotelian dialectic in 1527 and became the head of the Philippists, in whose circles mathematical subjects also were treated with great interest; as examples, see the editions by Joachim Camerarius (1500–1574) of the works of great mathematicians.

The French humanist school, founded by Jacques Lefèbre d'Etaples (1455?–1536), turned completely away from Aristotle. Its followers included Johannes Sturm (1507–1589), founder of the Protestant secondary school (*Gymnasium*) of Strassburg (1538) which later (1567) became an Academy with the right of awarding the highest degree in philosophy. The instruction in mathematics in the higher classes was implemented by the Latin editions of Euclid (1566) prepared by Conrad Dasypodius (1530?–1600) and the explanations by Bernard Barlaam (1564, 1572), by a general *Introduction to Mathematics* (1567–1596), and by two Greek-Latin *Dictionaries of Technical Terms in Mathematics* (1571–1573).

The passionate Pierre de la Ramée (Petrus Ramus; 1515–1572), a student of Sturm's during his stay in Paris (1529–36), labelled Aristotelian logic "a mere art of disputation" during his inauguration as master (1536), and in 1543 he developed his much fought over *logic* "on natural foundations." He was very interested in mathematical subjects, and he was the author of a Latin edition of Euclid (1545), of an *Arithmetic* (1555), and of a posthumous *Geometry* (1577). In *Scholae Mathematicae* (1569)—the introduction of which contains a historical survey strongly influenced by Proclus' *Commentary* on Euclid (Greek edition in 1533, Latin edition in 1560)— Ramée passed critical judgment on the Euclidean method, and while his criticism went too far on the whole, is was fully justified in many a detail. His pupil, Johannes Thomas Freige (1543–1583), elaborated on some points (1583). The desired simplification was accomplished, however, only under the renunciation of Euclidean rigor and of every proposition of somewhat greater difficulty. The ardent Lutheran Christoph Scheibler (1589–1653) adopted the same method.

But the leadership in the domain of pedagogy in the Early Baroque was not in the hands of the Ramists, but in those of the Jesuits. That order, founded by Ignatius Loyola (1491–1556) in 1534, established the *Collegium Romanum* in 1551, the *Collegium Germanicum* in 1552, the College of Ingolstadt in 1556, and in 1563–64 it took over the University of Dillingen (founded in 1549). The Jesuits furthered the mathematical disciplines with loving care, although they placed the emphasis not on theoretical research, but on practical, applicable knowledge. In keeping with the spirit of the age, the textbooks (like those of the Protestant schools) were primarily of encyclopedic character. Whoever wanted to enhance his knowledge, had the opportunity to educate himself on numerous editions—excellent for their day—of the classics, together with pertinent commentaries and supplements. An especially popular textbook was the compendium-like edition of Euclid (1574, and very frequently reprinted) of Christophorus Clavius (1537–1612) who had joined the order in 1555, pursued his studies of mathematics in Coimbra under Nuñez, and was appointed to the faculty of the *Collegium Germanicum* a short while later. The lectures and writings, revealing outstanding pedagogical skill, of the celebrated teacher supplied numerous highly talented and extraordinarily interested students with the intellectual tools to enable them to discover new details in the domain of infinitesimal mathematics.

The Jesuit schools taught philosophy along Neo-Scholastic lines; moreover, they applied the pedagogical principles of the Spaniard Juan Luis Vives (1492–1540), too. On the other hand, the Jesuits were uncompromising enemies of the Ramistic view and opposed ever more and more harshly the anti-Scholastic teachings of the Italian students of the laws of nature, such as Bernardino Telesio (1508–1588) and Galileo Galilei (1564–1642). They took relatively little notice of the views, more strongly inclined toward the realistic disciplines, of Wolfgang Ratke (1571–1635) and Johann Amos Comenius (1592–1670).

The interests of the Jesuits in natural science were centered very strongly on *mechanics*. They studied, above all, Commandino's editions of Archimedes (1558), Heron (1575) and Pappus (1565), then the treatise of Commandino on *centers of gravity* (1565), Monte's *Mechanics*, which was based on Heron (1577—special emphasis on the torsional moment), and the rather more profound investigations of Benedetti (1585: neglecting resistance, a heavy object falls no faster than a light one; rectilinear tangential motion). In the treatise *On the Center of Gravity* (1604) of Luca Valerio (1552–1608) they found the volume of the spherical segment,

$$\int_0^x y^2 \pi \, dx = r^2 \pi \cdot x - \frac{x^2 \pi \cdot x}{3} \text{ (with } y^2 = r^2 - x^2\text{),}$$

making use of the conically hollowed cylinder, the volume and center of gravity of the ellipsoid of revolution and of the hyperboloid of revolution of one sheet, $\int_0^x y^2 \pi \, dx$, using $y^2 = 2px \pm kx^2$; in the treatise of the same author *On the Quadrature of the Parabola* (1606) they found the layer-by-layer equality of centers of gravity of the hemisphere $\int_0^r xy^2\pi d$ and the parabolic

surface $\int_0^r t\,dx$ associated with it through $y^2 = r^2 - x^2 = rt$. In

the domain of *dynamics,* they advocated (in opposition to Aristotle) the Late Scholastic impetus theory (according to which a body thrown into a resisting medium is kept in motion by a force imparted to the body, which is slowly dissipated), and they were familiar with the exposition—conclusive for its time—by Dominicus Soto (1551) whereby free fall is to be regarded as uniformly accelerated motion.

The advent of Giambattista Benedetti (1530–1590) and Galileo Galilei (1564–1642)—both of whom started out from the impetus theory, but nevertheless manifested the opposition to Aristotle especially sharply—resulted in a tumultuous discussion, concerned at first less with the findings than with the methods of research. The new, ascending trend questioned the obligatoriness of the ontological outlook theretofore generally accepted as the basic view, and proposed to arrive at a quantitative description of the physical processes on the basis of experiment. Galilei's experiments relating to the motion of the pendulum, to free fall, to the projection of a missile and to motion on an inclined plane (Pisa, 1589–92; Padua, 1592–1600) led him to the law of inertia, to the graphic derivation of the distance-time curve for free fall from the velocity-time diagram, and to theoretical conceptions which approximate a special case of the law of the conservation of energy. His concise *Mechanics* (after 1604) was published (in French) in 1635; his more detailed exposition worked out on the basis of mathematical fundamentals (*Discorsi*) appeared in 1638. In consideration of the more extensive results of Valerio, he held back his *Studies on the Centers of Gravity* (1585), filled with the spirit of the "divine" Archimedes, and that work only became known after his death. Galilei took an ardent interest in the development of the infinitesimal methods all his life long, and again and again, he would spur his pupil, Cavalieri, author of the great *geometry of indivisibles,* on to a continuation of the arduous research (1621–29).

Johannes Kepler (1571–1630), enthusiastic emulator of the Neo-Platonists Pappus and Proclus, was guided in his mathematical activity by totally different points of view. In his profound *Weltharmonik* (1649) he analyzed at length the difficult Book X of Euclid's *Elements* and refuted Ramée's criticism (1569). There followed a number of contributions to the theory of the regular, semi-regular and star shaped solids, including the polygonal networks covering a plane without gaps, as well as contributions to the theory of musical intervals, to astrology and to astronomy (Kepler's Third Law). In his *Astronomia Nova* (1609) he established empirically the area theorem and the elliptical orbits of the planets; in this connection, he employed the focal point equation, $r = a + e \cos u$ ($u$ being the eccentric anomaly, determined by

$$\text{tg}\varphi = \frac{b \sin u}{e + a \cos u}\Big),$$

and on the basis of infinitesimal-geometrical considerations, lacking no more than the rigorous form of the dealing with limiting values, he arrived at

$$ct = \int_0^\varphi r^2 d\varphi = \int_0^u (a + e \cos u)\, du = b\,(au + e \sin u).$$

Thus, he was familiar basically with the parametric representation of the ellipse, which he had already approached closely in his comments on Witelo's *Optics* (1604: IX, 16) (fully carried out for the first time, by Mydorge, in 1631), and he knew also the relation $rd\varphi = bdu$. The transcendental problem named after him, calling for the determination of $u$ from a given $t$, was to acquire great significance in the Late Baroque (Wren, 1658; Gregory, 1668; Newton, 1676).

His *Doliometry* (1615) and its German supplement (1616) are the earliest scientific contributions to the art of sighting.

Kepler wrote for practical men and therefore used easy-to-understand rules of thumb which he made plausible on the basis of reasoning by analogy and succinct observations on the infinitesimal, without pressing on to rigorous proofs. This procedure was immediately rejected (1616) by Alexander Anderson (1582–1620?). Kepler's findings in this field, theretofore hardly touched, which were in some part correct and for the most part valuable as approximations at least—(volumes of the solids of revolution obtained by rotating segments of conic sections, the more difficult extreme value problems based on Pappus VII, 41) were the subjects of discussions in Italy, France and England; some points were confirmed by Briggs through direct numerical calculation (1625). Kepler's anticipation of the so-called "Guldin's theorem" and his use of curvilinear indivisibles were of great importance (integral substitution in stereometrical circumscribed forms).

Kepler's results exerted a certain influence on the Jesuit Bonaventura Cavalieri (1598?–1647) who endeavored to combine the ancient and Scholastic infinitesimal methods known to him with the new views of the Galileist school (publications from 1635 on). By letting, for instance, a plane "consist" of the totality of the parallel straight lines present within it, he was able to recognize as equal, areas which have equal sections between parallel lines, and to proceed analogously in three dimensions, too (the so-called "Cavalieri's Principle"). Also he used curvilinear indivisibles, beside rectilinear ones, for instance in the determination of areas for the spiral of Archimedes which he defined by means of $x = r, y = r\varphi$ as the image of the parabola $ay = x^2$. He amplified many of his individual results—to be sure, in cumbersome geometrical garb—also under a somewhat modified application of the Archimedean method of reasoning. Considered on the whole, his proof technique was still imperfect. His prettiest single result is the computation of areas of higher parabolas in the form

$$\int_0^a x^n dx = a^{n+1}/(n+1)$$

(1639); the pertinent proof was indicated, for $n$ up to 9, by

means of: $\int_0^a x^n dx = \frac{1}{2} \int_0^a \left\{ \left( \frac{a-x}{2} \right)^n + \left( \frac{a+x}{2} \right)^n \right\} dx$  in 1647.

Cavalieri was acquainted with the general approach to the determination of the coordinates of centers of gravity, and in dealing with surfaces he took into consideration the inhomogeneous distribution of mass, also. In 1632, he announced the modern method of computing the area of the spherical triangle; in 1639, he came out with the trigonometric treatment of quadratic equations and with the addition of logarithms. In his treatise on the *burning-mirror* (1632) he published the parabolic trajectory of a projectile without prior consent of Galilei, and this resulted in a temporary disturbance of the cordial relationship of the two scholars.

The Jesuit, Gregorius a S. Vincentio (Grégoire de Saint Vincent; 1584–1667), mathematically, the most talented pupil of Clavius, built directly on Archimedes and proceeded quite independently from Cavalieri in developing the so-called *ductus plani in planum*. It was a stereometric determination of

$\int_a^b yz dx$, where $y$ and $z$ are the co-ordinates of conic sections. In eminent conformity with the geometrical situation, he designated the concise, yet sufficiently explicit method of proof as *"exhaustion"*; this technical term was applied later also to the indirect Archimedean method which it would be better to call by some other name. Also Gregorius derived the spiral of Archimedes from the parabola through the introduction of curvilinear co-ordinates. We are indebted to him, furthermore, for the summation of the infinite geometric series on the basis of the proportionality of the series of differences to the original sequence, for the quadrature of the "virtual" parabolas $y = \sqrt{ax + b} + \sqrt{cx + d}$, for the cubature of the ungula (cylindrical "hoof") an oblíquely cut off cylinder whose cross section is a circle or a parabola, and for the division of the hyperbolic area $\int_a^b c^2 dx/x$ into $n$ equal parts by the interpolation of $n - 1$ ordinates between $y_0 = c^2/a$ and $y_n = c^2/b$ in geometric progression. His openly avowed espousal of the Copernican system of the universe made Gregorius suspect, and he had to change his place of residence frequently. He wrote his *Opus Geometricum* in Löwen in 1622–25, then studied it over critically in cooperation with Christoph Grienberger (1561–1636), the insignificant successor of Clavius, during his stay in Rome. It was almost ready for printing in 1629 when Gregorius suffered a stroke in Prague. The city fell to the Swedes, and the ailing Gregorius could just barely escape, leaving his papers behind; he got the manuscript back in Gent in 1641, but he took almost no part in the rather unsatisfactory final edition which came from the pens of his zealous students. The belated publication (1647) was, in effect, a work which by that time had been outdistanced in method. Its acceptance was prejudiced by the incorrect quadrature of the circle, from the earliest period of the activity of Gregorius, based on the erroneous equating of

$$\int_0^a \sqrt{a + x} \cdot \sqrt{a - x}\, dx \text{ and } \int_0^a \sqrt{a + x}\, dx \cdot \int_0^a \sqrt{a - x}\, dx.$$

This blunder had been a result of the purely verbal method of presentation which was difficult to follow through; it cost the author his scientific reputation in the eyes of his contemporaries; it was established in detail by Huygens in 1651 and by Vincent Léotaud (1595–1672), another Jesuit, in 1654. With this, the futile attempts at rescue by Gregorius's circle of pupils were finally abandoned. Of these attempts, only the one by A. A. de Sarasa is of importance. Here, de Sarasa stated that the hyperbolic area $\int_a^b c^2 dx/x$ was proportional to log $b/a$.

Charles de la Faille (1597–1652), one of the earliest pupils of Gregorius, developed further the latter's lost methods of determining the center of gravity. We owe him gratitude for a study, completed before 1628, on the center of gravity $(\xi, 0)$ of the circular sector $\int\limits_{-\varphi}^{+\varphi} \frac{a^2 d\varphi}{2}$ on a purely mechanical basis. His leading idea can be expressed infinitesimally in turning moments as follows:

$$\xi \cdot a^2\varphi + \frac{2a \cos \varphi}{3} a^2\Delta\varphi \approx (\xi + \Delta\xi) \cdot a^2 (\varphi + \Delta\varphi) \approx$$
$$\xi \cdot a^2\varphi + a^2 \cdot \Delta (\xi\varphi).$$

This expression yields, by affine transformation, the center of gravity of the general sector of the ellipse, and, by analysis, the center of gravity of the segment of the ellipse. La Faille showed then that the determination of the center of gravity and the quadrature of the segment of the ellipse were equivalent problems.

Valerio and la Faille were the authorities on whom Paul Guldin (1577–1643), a fellow-student of Gregorius a S. Vincentio based the direction of his work. He judged the methods and results of Kepler and Cavalieri as inferior; in 1640, he demonstrated certain special cases of the theorem named after him. The fact that this theorem had already been stated by Pappus was noted only by the revised edition, prepared by Carlo Manolesse (1660), of the *Collectiones* published by Commandino (1588). Guldin improved on la Faille's method in that he determined first the center of gravity of the arc and then that of the sector of the circle. But he was unsuccessful in his attempts at the rectification of the spiral of Archimedes and at the determination of the center of gravity of the surface of a solid of revolution, because in making use of a small obliquely lying versed sine he erroneously took only one component into consideration.

Another creative personality was Antoine de Lalovera (1600–1664) whose treatise on the quadrature of the segments of conics (1651) rested on mechanical fundamentals and was related in thought (but not in execution) to a contemporary writing by Huygens on the same subject. Both authors applied the method of computing the center of gravity based on

$$\xi \int_0^x y\,d\,x = \int_0^x x\,y\,d\,x,$$ where $y^2 = 2px \pm 2x^s$. The transition to a "quadratrix," $az = xy$, permitting the computation of the turning moment to be reduced to a pure quadrature, was typical of the purely geometrical point of view of Lalovera. A consideration involving three dimensions, which led him to the construction of the hyperboloid of revolution of one sheet and to the computation of the volume of that previously unknown solid, is deserving of particular attention. We must mention, finally, Andreas Tacquet (1612–1660), indirectly a pupil of Gregorius, for his assiduous study, dealing chiefly with the cubature of tori (1651, 1659).

In all these works, the Jesuits displayed excellent familiarity with the mathematicians of antiquity, who provided them both with inspiration and with methodological equipment. But they lacked the ability to establish and pursue a unified systematic science; they would lose themselves in obscure details, juxtaposed without sufficient inner ties. In their pursuit of the purity of the geometric method they passed over the algebra of literal notation, and they exhausted themselves in subtleties. This explains why their individual achievements failed to meet with well-deserved recognition; the distinguishing features of modern development were the pursuit of clarity, brevity, lucidity and system.

The Jesuits met with a similar fate also in relation to the second principal theme that presented itself in the domain of geometry in the early 17th century, namely the study of the conic sections. The decisive inspiration came from Apollonius, whose works were studied in Commandino's edition (1566).

Monte (1579) had already re-announced the forgotten paper-strip construction of the ellipse (first presented by Proclus, in his Euclid commentary). He also utilized the string construction of Anthemius of Tralles. Christoph Scheiner's compasses for drawing conic sections (compasses with a fixed angle at the handle, one oblique leg fixed and one adjustable leg), was made known in 1614, but it was of no practical use; the devices of Bramer (1634), Schooten (1646) and Christian Otter (1598–1660; posthumously announced instruments) could, however, be used very well for drawing conic sections and other curves. A particular interest in conics was shown by Kepler, whom we have to thank, for instance, for the first graphic representation of the systems $y^2 = 2px + cx^2$ (1609, 1615) and $y^2 = b^2 + 2px + cx^2$ (1615). A collection of theorems on conic sections, rich as to content but badly organized, was presented in Gregorius's *Opus Geometricum* (1622–29, printed in 1647) in which we find, for instance, the proof of numerous projective properties on the basis of metrical considerations. Cl. Mydorge's (1585–1647) theory of conic sections, in part very elegant, of which only the first four books appeared in print (of the others only the titles are known to us, from Mersenne's *Synopsis* [1644]), was less comprehensive and more strongly influenced by Apollonius. A collection of more than 1000 problems, some of them relating also to conics, was left behind by Mydorge after his death, but it still has not appeared in print; only excerpts of it have become known (1881, 1883).

The world-wide scientific activity of the Jesuits is demonstrated eloquently by the accomplishments of their missionaries in China in particular. For a long time they operated from the Portuguese settlement of Macao, founded in 1577, where Matteo Ricci arrived in 1582. As a result of his accurate forecasts of eclipses of the sun, which were still beyond the abilities of the native scholars, he was admitted to the court of Peking in 1601, and he gained the favor of the Emperor Shen-Tsung (1573–1619). In collaboration with the mandarins

Hsü Kuang-ching and Li Chi Tsao, he brought out a Chinese edition of the first six books of Euclid's *Elements,* based on the text edited by Clavius (1603–07). Johann Adam Schall, whose activity in China began in 1620, attained a position of even greater influence; in 1630, he was commissioned to carry out a calendar reform, and he rose to the post of director of the imperial observatory in Peking. His collaborator, J. N. Smogolenski, introduced the logarithms in China; as a result of the untiring literary and pedagogical efforts of his successors, the Fathers Ferdinand Verbiest (from 1659), Antoine Thomas (from 1685), Jean-François Gerbillon (from 1687) and Pierre Jartoux (from 1700)—all four held in high esteem by Shen-tsu —the most gifted native scholars freed themselves from the old tradition and espoused the mathematical-natural scientific views of the Occident. Their works were replete with the spirit of Euclid and the Cossists, and yet failed to produce any lasting effect because the China Mission was disavowed by Pope Clement XI in the controversy concerning rites (1715) and its activities had to come to a halt, with the result that the lively exchange of ideas between Europe and the Far East was lost again.

The Early Baroque brought noteworthy advances in every special field of mathematics explored in that period; in the form of practically applicable approaches and methods, it brought a far-reaching rise in the level of general education, but it still lacked the most important tools, namely: a unified system to set the single achievements in their proper places, and a rigorous, generally valid method of proof. Leading personalities were well aware of this lack. Agrippa von Nettesheim had advocated scientific scepticism in 1527. It was completely in the spirit of this viewpoint that writings of Sextus Empiricus were being published (1562, 1621). The *Essais* (1580 and thereafter) of Michel Montaigne (1533–1592), displaying the same trends, were written in an epoch teeming with exuberant intellectual energy that could be kept within bounds only by laborious effort. Consequently, their keynote was not resignation, but a voluntary acquiescence in the divine order manifesting itself above all in the laws of nature and natural processes that are independent of all human volition. This was the creative approach that was to produce new possibilities in the High Baroque (Descartes). An equally great effect was produced by Francis Bacon (1561–1626) who developed a natural-scientific program for the future, that was to fascinate the next generation. This effect was produced irrespective of the fact that he failed to appraise correctly the decisive advances of his contemporaries in the domain of natural science (Copernicus, Galilei, Kepler) and that in his own works he did not rise above very amateurish approaches. According to his views, Bacon still belonged to the Early Baroque period; according to his cherished aims, he already belonged to the High Baroque, of which period he was the herald.

# Index

129